Fertility from the Ocean Deep

Fertility from the Ocean Deep

by Charles Walters

Acres U.S.A.
Austin, Texas

Fertility from the Ocean Deep

Acres U.S.A.
P.O. Box 301209
Austin, Texas 78703 U.S.A.
(512) 892-4400 • fax (512) 892-4448
info@acresusa.com • www.acresusa.com

Printed in the United States of America

Publisher's Cataloging-in-Publication

Walters, Charles, 1926-2009
Fertility from the ocean deep / Charles Walters. Austin, TX, ACRES U.S.A., 2012.

    xii, 180 pp., 23 cm., charts, tables.
    Includes index.
    ISBN: 978-0-911311-79-2

    1. Trace elements in agriculture. 2. Trace elements in plant nutrition. 3. Trace elements in animal nutrition.
    4. Micronutrient fertilizers. I. Walters, Charles, 1926-2009 II. Title.

S587.5.T7   631.81

Contents

Foreword vii

1 The Missing Link to Health 1

2 The Nutritional Center of Gravity 11

3 Ocean Traces, Ocean Solids 23

4 Stepping into the Light 33

5 The *In Vivo* Factor 49

6 Seaponics Farming 67

Photo Interlude 79

7 The Ocean Presides 87

8 Sea Energy Agriculture 101

9 The Ionic Ocean Water Connection 111

10 A Sea Energy Testament 119

11 The Bottom Line 135

12 The Forgiveness of Nature 141

13 A Pantheon of Minerals 151

Afterword 165

Sources 168

Index 169

Dedicated to the memory of Maynard Murray, M.D.
Pioneer, Researcher, Seeker

Foreword

As a child of the high plains, I did not see the ocean until 1944 from the shores of Biloxi, Mississippi, where the Army Air Corps had sent me for flight training. This view was enhanced when I hitched a ride on a flight over Cuba in a Navy PBY. But it was not until much later that I encountered the ocean in a meaningful sense, this time on the Colorado plateau about a mile above the sea. I was assisting Al Look in the publication of his book, *1,000 Million Years on the Colorado Plateau, Land of Uranium*. Look had discovered a fossil the paleontologist allowed him to name *Sparactolambda looki*. Look was a journalist by profession and an amateur paleontologist on the side. He once told me it was the function of the amateur and the writer to buy the books of the credentialed professors, honoring them on the side.

It was *Sparactolambda looki* that really caught my attention on the high plateau. Stamped into sedimentary rocks and soil layers were carbon copies of ancient sea life, including various species of fish and other ocean creatures. Well over 100 million years ago, explained Look, the ocean covered the Colorado plateau, enriched its soil and placed on deposit the nutrients biological life now demands.

In 1672, Robert Hook published a book of drawings depicting fossils preserved in stone, many of them so small they required the use of the newly invented microscope. Hook noted that many fossils did not resemble any form known to be alive at that time. He argued that some species must have become extinct. This changing of species was used to date the strata in which they appeared. Hook was ignored much as proponents of sea energy agriculture are often ignored today.

Hook added the Earth's evolution, cooling, crumbling and decay. Had he lived longer he might have discovered this crumbling as a precursor to the manufacture of ocean energy.

I finished my book proofing for Look and went my own way as a reporter for the *Rodeo Sports News,* and for all the world the ocean might as well have washed over this experience the way it washed over the *Pequod* on the final page of *Moby Dick.* Not long after founding *Acres U.S.A.* in 1971, I had the occasion to be in Rome on other business. The Villa Banfi people helicoptered me and my wife, Ann, to their northern Italy vineyards for a wine journal assignment. There the ground shook as heavy machinery hammered reinforced concrete stakes into the hard sediment of a mountain that once lay undersea. The clues could not have been clearer. Here were the life forms I had encountered on the Colorado plateau stamped in stone with all the grandeur of Caesar's image on coins. The ocean had been there before, and geologists used to tell you about when and for how long. I can't say a light bulb went on in my mind, but in fact Italy's vineyards asked a question that remained unanswered until the day Edmund Hillary and his native guide Tenzing Norgay climbed Mount Everest.

Mount Everest is situated behind the Himalayas. It separates the Tibetan Plateau from the Indian Plains below. The giant mountain was first measured from a distance and pronounced the highest land on planet Earth. It was named in 1856 after Sir George Everest, the Surveyor General of British India. Hillary and Norgay reached the summit of Mount Everest in 1953.

That was also the year James Watson and Francis Crick revealed to the world the structure of DNA. The connections that tie these events to one another will emerge as our *Fertility From the Ocean Deep* story unfolds.

The conquest of Everest was an achievement that vied with the coronation of Queen Elizabeth II for attention. It may have been the last great earthly adventure, a symbolic stepping stone into space.

It was and still remains quite a mystery, a geographical question. Under the sponsorship of the Royal Geographical Society, the 1953 expedition repaired the mistakes of some five earlier failures and added to the accumulation of knowledge about the ascent.

Near the top, the two climbers encountered a great slab of rock. It was not difficult to traverse since it exhibited deep scars, if such a term can be used. As the ice axes created footholds and the climbers crossed the last obstacle to the top, both noted the impressions of shells and fish, creatures perhaps not unlike those I encountered in Colorado and Italy. Apparently, the ocean had already been there, one way or another.

When *National Geographic* finally detailed the ascent of Everest, my thoughts went back to the Bible and to the allegory or metaphor of a great flood that occupied the Earth. Whatever the floods of Gilgamesh or Genesis have to do with land masses rising out of the sea is anyone's guess. One thing is certain, however: there isn't a spot on planet Earth that hasn't been covered with ocean water at one time or another.

Come walk with me under that ocean, as if all its waters had been vaporized and held in escrow pending the completion of our trip. Listen to the winds as they dive into the deep night and day and hear the roar of volcanoes as they spew out their treasures for the digestive pleasure of the ocean water. Observe the fossils that open a crevice into the center of the Earth itself. Chart the mountain ranges from trenches as deep as Everest is tall. They rise from the ocean at times and then disappear again when water is in its place. As you walk the Pacific canyon with Paul Bunyan strides, you'll see treasures beyond the dreams of avarice. As you follow the mountain ranges across South America, then into the Atlantic areas, and then into Africa, you realize the Alps were once situated to the south of the Mediterranean, and indeed Everest once rose from the sea the way the Phoenix rose from the ashes.

Now put all that water back in its place, having realized that exploration of the ocean floor is much more difficult than a trip to the moon. Now stand on some spit that juts into the Atlantic or Pacific or any of the oceans of this watery planet. Watch the waves slide in or out of the shore with sonic force. Watch the water eat away granite, soften shorelines with sand, then attack the land with a vengeance.

The ocean is always in turmoil, as is life itself. There are currents that distribute heat and food and life as they push that life toward the Poles. The earth beneath the sea is rich in essential ingredients – lime, potash, phosphates, trace minerals. These nutrients are leached out of older soils, delivered via volcanism, tamed and mixed by the ocean in a complex process that makes the ocean the nutritional center of gravity. There was a time some years back when experts believed as few as 10,000 volcanoes were located in the ocean depths. When they delivered their potent mineral load, the waters boiled and then moved out to mix with the grand miscellany of waters worldwide. That number has now been enlarged fourfold, according to Robert Felix's *Not by Fire but Ice*. Ancient delivery of the nutritional richness, much like life itself, came from the sea.

Volcanic soils such as those used to grow coffee enjoy the richness of soils located by the sea, a consequence of land eruptions. I have seen the coffee groves of Costa Rica and Brazil and marveled at their survival. The residue once covered by volcanic ash suggests the role of ocean dirt for complexed and depleted soils.

These few details are set down here to prepare the stage for a detailed look at ocean energy, ocean-grown vegetables, and wheat and rye grasses, with a hint at the possible application of ocean energy to row crop and pasture production.

There is a little book French children use as classroom fare that makes a point for our time. It is called *The 29th Day,* and it is used to teach children about exponential growth. A lily pond, the riddle has it, contains a single leaf. Each day the number of leaves doubles – two leaves the second day, four the next, then eight, then sixteen, and so on. If the pond is full on the 30th day, the question has it, at what point is it half full? The answer is the 29th day.

Without knowing it, American agriculture has reached the 29th day. Its lands have become exhausted of carbon, depleted by N, P and K salt fertilizers and poisoned by toxic rescue chemistry. The inevitability of so-called scientific practices has delivered degenerative metabolic diseases to a population that is also expanding exponentially. The only remedy for a nutritional shortfall in lands and even hydroponics is the ocean itself.

In 1976 I invited Dr. Maynard Murray to speak at the annual Acres U.S.A. conference. Almost every word he delivered at the conference is presented in the pages that follow. Maynard Murray's 50-year stint at exploring the miracles of sea energy agriculture did not die with that great physician and scientist. They were refined and sustained by Don Jansen, whom you will also meet in this book.

During a lengthy conversation, I asked Dr. Murray about Mount Everest. Did the ocean really cover the world's highest mountain at one time? "Yes, of course the ocean covered the mountain before it rose from the sea," was his answer. "All life started in the sea, the life that is now dependent on plants. The degeneration due to immoderate nutrition bodes ill for the foods we grow, and I have no doubt that the single cause for all disease conditions is malnutrition." The quote must be recalled from memory because the original note has vanished. When I related the quote to Alwyne Pilsworth of England, he responded as follows:

"All life begins in the sea, but all life on Earth is sustained by food from plants. Dr. Maynard Murray refers to the present level of disease and illness in the world as being due to unbalanced and overfeeding of plants. This is fundamentally sound. To enable us to benefit from such an idea, we must learn how to produce food by a system that is sustainable. Life on earth can only be made sustainable by using Dr. Carl Oppenheimer's bugs to clean up the root environment and deposit all available organic matter adequately, prior to returning it to the soil in the form of well-made compost."

The pages to come will describe Dr. Maynard Murray and follow his exciting career, as well as the career of an idea – an idea that he and his cohort found applicable in a tough commercial world. Using ocean water first concentrated, then diluted for pastures, row crops, vegetables and fruit trees appears to be an idea whose time has come.

Whether growers accept this element of nature's gifts, the primeval sway of the ocean will go on, uncaring. There will always be rollers on the beach, high sea levels gravitating to the northern and southern latitudes, mammals traveling in pods, the pull of the moon stirring the great cauldron so that tides are forever indebted to it and the song of the stars.

1

The Missing Link to Health

The man who came down to the wharf each day was a physician on the hunt for tranquility. His specialty was ears, nose and throat, but his interests ranged across a broad spectrum. When he made his rounds at the hospital in Boston, he grew increasingly appalled by the institution's caseload of patents suffering from cancer, diabetes, arthritis, osteoporosis, arteriosclerosis and other forms of degenerative metabolic disorders, all diseases of high civilization.

Dr. Maynard Murray, M.D., had taken his medical training at the University of Cincinnati in the early 1930s while working at a physiological laboratory, and there he spent a decade testing the art and science of his profession. Even then this great man was engaged in finding an answer to two modern scourges, cancer and arteriosclerosis.

A short biography written the week Maynard Murray retired at age 71 called him "a voraciously curious man by nature." It was more an understatement than a studied analysis, for Maynard Murray was a medical carbon copy of an inventor who had also retired to the same Florida neighborhood, Thomas A. Edison. Maynard Murray's voracious curiosity led him to file and hold some 50 patents before he passed from the scene.

Fifty years intervened between that walk to the wharf and immortality, and they were marked by an epiphany, a rebirth from the certainty of American Medical Association (AMA) medicine to the mysteries of the ocean.

We came from the ocean, he often told interested listeners, and now we reach for the stars. We do not know what brought the physician to the waterfront for badly needed relaxation. Most of his papers have vanished, except for highly individual reports in the professional literature and patent applications on file in the government office that handles such matters.

Ed Heine, an Illinois farmer who worked with Murray for 30 years, tells us Murray had become convinced that the human race really was what it ate, and that the quality of that fare was deteriorating even while achieving cosmetic beauty.

The soundness of his reasoning was firmly anchored in a debate that made Louis Pasteur the bitter enemy of one Antoine Beauchamp. Pasteur thought most human misery was caused by germs, literally copies of the evil spirits Martin Luther knew were everywhere. Annihilation of these microorganisms was devoutly to be wished. Antoine Béchamp said no, no, these organisms are a part of nature, much like the *E. coli* in the gut of warm-blooded animals. It is the human terrain, he said, the bodily milieu, that governs human health. Allow a weakness in the immune system to develop, and then opportunistic organisms take over.

Pasteur won the debate, and as a result the world goes blithely about the business of pasteurizing milk, cooking food to death, irradiating meat, and microwaving food to inhibit the growth of bacteria, often beneficial bacteria.

Did these things cloud Maynard Murray's mind as he sat down without pencil or pad or even reading material? The tired body takes over completely when waves gently slap the pillars that hug the floor of the sea. The act of occupying a deck chair tends to lure the human machine into the primeval rhythm of the ocean. A spell takes hold and fine-tunes the mind. Thoughts flow with the cadences of rollers on the beach, waves in the tides, and the seagulls as they hunt for food and pester fishermen.

A roller comes in. It erases marks on the beach. At some uncertain point, the mind comes to life again. Unburdened thinking time always serves up new connections.

Nature reveals her secrets slowly and only to willing listeners. This business of empty foods was a matter of record. As early as the 1930s, it was an article of faith in the United States Department of Agriculture that there was no relation between the vitamin content of foods and the chemical composition of the soil in which they are grown. Some years later these precise words were entered into the *Federal Register,* and the Department of Agriculture and the Food and Drug Administration have held to this position ever since. Once upon a time it may have been true, or it might have been true in a few select situations, farm to farm, where growers hadn't heard too much about nitrogen, phosphorus and potassium (N, P and K) and toxic genetic chemicals, and had not yet learned to pursue the ruthless rape of the soil.

Farming was changing rapidly when Maynard Murray took the position that the common denominator for degeneration was malnutrition. Small farms were becoming bigger farms, and bigger farms were becoming great consumers of acidulated salt fertilizers. Moreover, certain types of vegetable and fruit production had become concentrated in three or four states under harsh monoculture conditions. The fact was that mineral and vitamin composition of food crops varied widely, depending on soils in which they were grown and how they were fertilized.

This much was discovered as early as 1942, when the Federal Security Administration, working under the National Nutrition Plan, took steps to sustain nutrition levels despite the exigencies of war.

The variations found in this study should be at once apparent. Instead, these findings prompted Dr. Firman Bear's Rutgers University study in the late 1940s, a 10-area look at variations in the mineral composition of vegetables. Bear's findings were made a matter of record in *The Science Society Proceedings of 1948.* Variations in food minerals turned out to be wide and dramatic. The iron content of tomatoes, for instance, varied by over

1,900 percent, depending on the chemical composition of the soil in which they were grown.

The record of five vegetables in the Bear study exhibited less dramatic, albeit terribly significant, variations for soybeans, cabbage, lettuce, and spinach. There were additional studies in Michigan, New York, Canada, California and Florida, all damning in their finality.

Bear's adversaries, William A. Albrecht, Cyrus Hopkins, Louis Bromfield, and dozens more, all members of Friends of the Land, had been right all along. Worn-out soils produced worn-out men and women in the fullness of time.

That *Federal Register* entry a few years later exhibited the reluctance with which the FDA and the Department of Agriculture face any change in thinking: "Scientifically, it is inaccurate to state that the quality of soil in the U.S. causes abnormally low concentrations of vitamins or minerals in the food supply produced in this country."

Such a retreat from reason can have only one origin. The economic mandate of agribusiness requires the use of distorted accounting principles to justify demineralized, devitaminized and debased foods.

Indeed, the 1959 *Yearbook of Agriculture* blindly proclaimed what to officialdom had become settled science: "Lack of fertilizer may reduce the yield of a crop, but not the amount of nutrients in the food produced."

If the ghosts of drowned sailors had arrived at Maynard Murray's observation pier and whispered, "This is nonsense, insanity, a retreat from reason," it would not have been phantoms speaking, but nature's own reality. As a scientist, he knew that calcium, nitrogen, phosphorus and a long list of other elements – including the trace nutrients – were required to synthesize amino acids, proteins, vitamins, enzymes, lipids, octocosanols and phosphatides: the building blocks of plant food life. Plants and microbes, even those in the cow's gut, synthesize the amino acids that make up proteins. "Both plants and animals assemble their proteins to provide the reproductive functions," the great soil scientist William A. Albrecht wrote. "These are the only compounds through which the stream of life can flow."

The belief that this complicated life process can be serviced with simple N, P and K fertility harks back to Justus von Liebig, but it surfaced anew hard on the heels of World War II. At that time, "two false premises swept the republic of learning," wrote Sir Albert Howard, "partial and unbalanced fertilization and toxic rescue chemistry."

Ignored or left out entirely was consideration of the trace nutrients. They had been left out before. When glaciers scraped their way across North America as far south as the northern edge of Arkansas, they redistributed the very minerals oceans had gifted those areas before. The trace minerals available to most farmers are complexed, that is, rendered inaccessible to plant roots. Their soil is littered with set forms of nitrogen, phosphorus and potassium (N, P and K) even if they haven't been farmed out. Since then, distorted accounting procedures and the breeding of corn that can live on less has accounted for more bins and bushels, and less quality.

In terms of mineral uptake, open-pollinated corn still has an enviable record. The late Adolph Steinbrown of Fairbanks, Iowa, put the matter in perspective by having samples of corn tested for ingredients sometimes added to commercial feeds. One was a hybrid he'd grown. The other was OP corn. The OP corn grown on his unfertilized glacial soils contained 19 percent more crude protein, 35 percent more digestible protein, 60 percent more copper, 27 percent more iron and 25 percent more manganese. Compared to some 4,000 samples of corn tested in 10 Midwest states in a single year, Steinbrahn's open pollinated corn contained 75 percent more crude protein, 875 percent more copper, 345 percent more iron and 205 percent more manganese. The same trend was also seen in the content of calcium, sodium, magnesium and zinc. It can therefore be said that OP corn can contain an average of over 400 percent more of these nutrients for which tests were made.

Another farmer of the same era, Ernest M. Halbleib of McNabb, Illinois, confirmed the failure of hybrids to uptake certain mineral nutrients from even an organic farm where the traces were available in glacial abundance. Comparing Krug OP corn and a standard hybrid in the laboratories of the Armour

Institute of Research, Chicago, spectrographic testing revealed the hybrid to be short of nine tested minerals. The hybrids failed to pick up cobalt and other traces. Both varieties had the same chance to pick up a balanced ration. "The reason I mention cobalt," wrote Halbleib, "is that we found on the 16 farms tested that no hybrid picked up cobalt, and in all the tests the hybrid was short seven to nine minerals, always exhibiting a failure to pick up cobalt." The core of vitamin B$_{12}$ is cobalt. Dr. Ira Allison of Springfield, Missouri, found cobalt implicated as a cause of brucellosis and undulant fever, and cobalt is part of the cure. Hybrid corn merely works for farming by producing bins and bushels, but without the nutrient load good health requires, corn's reason for being. On most conventional farms, the traces can't be picked up because they aren't there, and OP corn can't grow at all.

Early in the first decade of the last century, the New Jersey laboratory of George H. Earp-Thomas discovered the near total absence of cobalt in American soils. No art of agronomy seemed capable of restoring this essential nutrient. This shortage soon had laboratories off and running, measuring the effect of trace nutrients on enzyme production. Not one could be produced in the body without mineral assistance. Maynard Murray once confided to an associate that the rate of enzyme discovery is slowed by the pace of trace mineral evaluation, and some 400 years will be required to get the full picture if the present rate of discovery is not hastened. All enzymes require trace mineral keys. Dr. Jerry Orlasch has called them "the spark plugs of the body." It was Earp-Thomas who cemented into place the proposition that a body with high mineral content could remain disease free. With this mineral assistance, the body could maintain a fluid electrolyte solution in its cells, osmotic equilibrium or pressure. His observation seemed to confirm Antoine Béchamp's belief that if the terrain was not suitable for disease, disease couldn't happen. The syllogism concluded its premises. Fluid and water into and out of cells governed blood pressure, kidney function, nerve process and hydration.

By the middle of the 20th century, the trace minerals had almost become a discipline, but the studies seemed to deal with them one at a time. André Voisin's *Soil, Grass and Cancer* dealt largely with copper and zinc. Disease conditions started being tagged accordingly to an average or a shortage of this or that nutrient, macro or trace. New paradigms were installed in the literature. Trace minerals were said to be under homeostatic control. That is, they obeyed a law of homeostasis. They were essential. Essential did not mean more was better. Above the curve, they could be toxic, even lethal. First a dozen elements became characterized as essential. As new data were assembled and a way was found to market these essentials, the number became enlarged. Each decade a new physiological role for still another trace nutrient is identified. Unless the pace of discovery is picked up, the full Mendeleev table will not be understood for several centuries.

Maynard Murray pondered these facts in his talk before an Acres U.S.A. conference in 1976. Human beings, he said, can use the essential elements only in shielded form. If the job of complexing them isn't handled by plants, then it is up to bacteria in the gut to come to the rescue. Obviously, this is impossible if there is a great overload of this toxicity. The mystery of salt comes to mind whenever ocean spray delivers the water's taste to human beings. The Chinese once executed criminals by forcing them to consume several tablespoons of salt. Yet the same salt is used at mealtime around the world by toddlers and grownups alike. Physicians recognize this and prescribe salt-free diets for pregnant women and men with heart problems. Yet these diets never omit the claim that a celery stalk has as much sodium chloride in it as a typical shake of salt over a plate of peas or beets. There is a reason why the sodium chloride in celery is not toxic. Only sodium chloride in the inorganic state produces toxic effects. Withal, sodium remains pretty much a mystery, yet it is one Maynard Murray listed as hypothetically solved. The metes and bounds of that solution anoint every step along his career's way.

We are never done with mysteries when it comes to nature. Aluminum is a good "for instance." It seems to play an important role in some few conditions and turns up permanently in green leaves, thus the ongoing speculation that there may be an association between aluminum and chlorophyll.

Whether sodium could be substituted for potassium was investigated as early as World War II, and the topic received maximum attention from Maynard Murray as he watched fishermen unload their catch. Conversations with men who had spent a lifetime invading a veritable Serengeti in the water led Murray to perform necropsy examinations on various fish. He couldn't recall seeing such pristine livers, spleens, organs of every type.

"Are there ever any tumors?" he asked the friendly fishermen. "Not that we've ever seen," was the usual answer.

Sodium a substitute for potassium? The very idea suddenly elevated a perceived ocean contaminant to nutrient status. Could sodium take on the task of potassium in assimilation? It could! Maynard Murray's rest on the wharf turned into an intellectual feast. In crop production, such a substitution would allow more of the potassium to function in the seed. This would mean that sodium needs to have potassium available to it, and that it tends to preserve soil calcium, magnesium and potassium. It could then assist in plant nutrition when soil potassium was not sufficient for the requirements of the crop.

Rivers were full of fish with tumors not unlike those Murray saw among hospital patients. Even though these rivers drained toxic soils and spilled their contents into the ocean, the man-made molecules hadn't afflicted ocean life with tumors, at least not in the early 1950s.

In 1949, the U.S. government declared that the nutrient in least supply limited plant growth. In a manner of speaking, there isn't such a thing as a major or minor nutrient. All are critical in their assigned roles.

Maynard Murray opened up still more fish. He knew the government had set up Poison Control Centers nationwide. This step was made mandatory as the two false premises, imbalanced salt fertilizers and toxic rescue chemistry, swept those republics of learning of which the great agriculture thinker Sir Albert Howard wrote. It was an amateur agriculture coming on, one that failed to realize that a true anatomy of weed, insect, bacteria, virus and fungal control is seated in fertility management, including trace mineral management. If the Friends of the Land knew that plants in Iowa with suitable loads of exchangeable nutrients could protect themselves, it was knowledgeable everywhere. To disturb the soil's microflora — algae, fungi, protozoa — was to invite degenerative diseases, even if the farm product looked beautiful.

If science has failed to synthesize the chemical and biological energies of enzymes, it has succeeded in inserting pollution into soils, plants, animals and human beings. As we unveil the story of Maynard Murray's quest, we will examine anew the ocean's key to enzyme production. For now it is enough to concentrate on civilization's assault on the cell, this route to mischief and the anomaly called cancer.

Cellular damage due to malnutrition or the invasion of toxicity can cost a farmer all or part of his crop. The same damage in human beings can cost a nation its heritage and its future. Damage to the sperm or ova in a human being can cause malformation or mental retardation in future generations. It can also contribute to degenerative metabolic disease.

In every cell — plant, animal or human — there are chromosomes, which carry almost all of the information needed to direct the cell's growth, division and production of chemicals such as proteins. These chromosomes are composed of information-bearing genes.

Farm chemicals are radiomimetic, meaning they ape the character of radiation. Radiation itself can damage the chromosomes either by altering the character of a single gene so that the gene conveys improper information, called "point mutation," or by actually breaking the chromosome, called "deletion."

The cell may be killed or it may continue to live, sometimes reproducing the induced error. Some types of cell damage cause genetic misinformation that leads to uncontrolled cellular growth, cancer.

Listen to Dr. Maynard Murray, speaking near the end of his career:

"There can be no life without a transfer of electrical energy. Each cell is a little battery. It is capable of and does in fact put out a current. If it is unable to put out a current, the cell is dead and can never return to living tissue. Anything living alters its environment for its benefit in order that it may live and reproduce.

"Life is contained in a cell. It is surrounded by a definite volume. Cells vary in size. The largest cell on Earth is an ostrich egg. The smallest cell is a bacteria. In warm-blooded animals, the reproductive cells are both the largest and the smallest. The sperm cell is the smallest, and the egg cell is the largest. The cells are able to carry on the process of life. They need only food from the outside, they can manufacture many of their requirements, they can break down complex compounds and synthesize their own body tissues.

"A virus cannot do this. It has to live within the cell as a parasite. Living tissue has to get its food by either concentrating or diluting its environment in order to make that environment a part of its tissue.

"All of life is parasitic. One living thing lives on another. The exception is plant life, which contains chlorophyll. Using colored dyes, plants can synthesize living tissue out of inorganic materials. The pigment in the retina of the human eye enables the human animals, with the aid of light, to synthesize food, proteins, etc., out of simple inorganic components."

The syllogistic conclusion became the focal point of his report and had been the focal point of his life's work. "The green plants," he said, "will not use organic materials."

Organic farmers, he said, were doing the right thing, but they were using the wrong name. The plant uses inorganic materials and makes them organic. Plants take inorganic nutrients, animals do not; they need organic food.

It was a thesis he had spent nearly 30 years refining while on the hunt for trace nutrients, the missing link to health.

Ed Heine of Illinois recalls Maynard Murray's oral biography exactly as he heard it. His days of observing the fish catches as they were unloaded, and his tryst with the fishermen came to an early end, as do all days and hours away from work. A few months intervened before Maynard Murray took a leave of absence from hospital rounds and went to sea on a trawler for eight months.

He wanted to learn about oceans and the origins and destiny of life.

2

The Nutritional
Center of Gravity

"Remember thou art dust and unto dust thou shalt return."

No one who hasn't spent a life in solitary confinement could have failed to hear that biblical injunction at one time or another. As suggested in the previous chapter, the dust for human cells is the dust of the soil, and now Maynard Murray must have wondered whether the traces in ocean water were not superior. He set out on his eight-month sea odyssey to find out. The parable contained in the biblical saying required meditation because it was more than a philosophical saying. It was a scientific truth, wrote André Voisin, "which should be engraved above the entrance of every facility of medicine throughout the world. We might better remember that our cells are made up of mineral elements which are to be found at any given moment in the soil of Normandy, etc." He did not add "in the water of the ocean" because he wrote long before Maynard Murray followed his quest. If these dusts have been improperly assembled in plant, animal or human being cells, Voisin held "the result will be imperfect functioning of the latter."

Of all the living organisms that present the photograph of their environment in their makeup, none seemed to elude degeneration like ocean life. Land-based health specialists knew that dusts of the soil, meaning minerals in trace form, govern the presence or absence of good health. Long before metabolism and enzymatic functions were known, a farmer knew his animals were strictly a product of the soil. In one of the matchless scientific papers he left behind when he passed from the scene in 1973, William A. Albrecht called attention to the dwarfing of the classic Percheron within a few generations after being removed from their French soil of Perch-Garrison. He assigned the same fate to the badly fed Madrasi tribe of India.

"We are what we eat," noted Maynard Murray in speeches to groups interested in food and agriculture. Malnutrition finds its common denominator in the absence of a few trace minerals from cells.

Billions of species live on the Earth, most of them still unnamed. An estimated 90 percent live in the sea, the probable origin of man himself. With cancer taking perhaps one in four circa 1950, and no discernible cancers on the other side of the equals sign, there was more mystery to the ocean waters than the underwater finds Jacques Cousteau could expose on film.

One of them is the chemical composition of water. Scientists and aquarium owners had been trying to synthesize ocean water for years. They followed nature's blueprint to the letter, and yet ocean fish species cannot live in the manmade seawater. This inability to create seawater in the laboratory was a puzzlement until it was noted that four elements of the ocean were present in the bodies of microorganisms — cadmium, chromium, cobalt and tin. All these have been found in the remains of ocean organisms. Food hopes notwithstanding, scientists cannot synthesize life. Dead plus dead equals dead. It takes life to create life. Those microorganisms that contribute nutrients are life.

Before he boarded that trawler for a long sabbatical away from the hospital wards, Maynard Murray read the literature on the world's oceans. Seaweed figured in his thinking even before his first stint of relaxation at the ocean's edge. Seaweed, he knew, contained essential microorganisms.

It was quite possible that the seaweed organisms were the most essential part of the seaweed connection. Seaweed was being used in every research laboratory and in every hospital. Throat cultures are put in agar, which is nothing but gelatin in seaweed. Seaweed has proved to be the most ideal nutrition for almost all bacteria. There is a reason for this. Bacteria refuse inorganic nutrients, which makes seaweed on the petrie dish a perfect balance for culture purposes.

After all, bacteria are basically aquatic animals. They live in water and they find their pool of water in the cell while on the hunt for sustenance. The evidence was out there, beyond the places where continents jutted spits and elbows in the path of a living ocean. Maynard Murray knew he would feel the wind and spray in his face, the roar of the ocean in his ears. He would see it all from the surface. His fishermen friends would bring aboard denizens of the ocean they had fished. A cure for cancer was his passion, his prime interest.

Necropsy examinations on the wharf told Murray that there was no cancer, no hardening of the arteries, no arthritis in the sea creatures he dissected. Why? That was a question no one could answer, probably because it wasn't asked.

Even before he boarded that trawler for an eight-month ocean stay, his reading and research brought in the mother lode. He found river trout that had developed cancer of the liver. Salt-water trout species did not. Why did indigenous Innuit people seldom suffer cancer or coronaries? At least they did not until traders brought in processed and debased foods. Weston A. Price's classic book *Nutrition and Degeneration* brought this man of modern medicine up to speed. All the world over, degeneration followed mineral-deficient food the way death follows cancer. Modern medicine taught Maynard Murray that fats and salts inserted into the diet should be held to a minimum in order to avoid arterial and heart problems. Yet the same Innuit violated AMA rules when they ate blubber from whales, all of it saturated fat. Warm-blooded whales and cold-blooded sea creatures alike live in a saltwater environment and consume saltwater daily. Why was it that so many mammals on land developed cancer, whereas mammals of the sea did not?

Traveling around the world, Maynard Murray dissected whales in Peru, seals in the Pribilof Islands, manatees in Australia, and fish species everywhere. It took him 20 years not to get all the answers, but to learn most of the questions. "He examined the arterial systems of thousands of sea mammals — autopsying 50- and 60-year-old whales" — this according to one of the last journalists who interviewed him. "He consistently found they had the arterial systems of newborn calves; clean, healthy, none with the signs of deterioration humans suffer."

Why no arteriosclerosis among warm-blooded mammals carrying tons of fat and living in salt water? Why did tissue samples sent to a pathology laboratory exhibit no signs and symptoms of malignancy? On the other hand, why did land animals develop both?

The ocean was a great teacher. It was impossible to enter the ocean without encountering waters that arrived from at least as far as the Equator days ago. Some of that water journeyed from Antarctica, some 9,000 miles distant.

The sailors taught him that surface waters meander around the oceans in gigantic water ponds. Some of that water takes centuries to move along its assigned course, while some of it swims atop the ocean so swiftly it can retard the outbound passage and then cancel the lost speed on the trip back home.

The ocean does not allow water to stagnate. The moon does its job keeping ocean waters stirred up. But ocean tides are not the whole story.

The Earth itself is a unique planet. It is most certainly the only living system in our solar system, assuming no one turns up bacteria or fungus on Mars. As planet Earth swings around the sun in a somewhat eliptical orbit, it turns on an imaginary axis so as to toast the tropics and refrigerate the poles. Heat expands equatorial surface waters, which then flow downhill to both the North and South Poles.

On the floor of the ocean, the water is always just above freezing. There is always an exchange, something akin to salt water flowing through the Bosphorus Strait, fresh waters from Siberia's rivers flowing out into the Mediterranean, salt water flowing into the Black Sea.

Other forces figure in sustaining the ocean mix. Some are understood, some are not. The Earth spins at 1,000 miles per hour. As it tries to spin out from under its covering of water, it spins to the east. Therefore waters hammer the western shores.

This spin causes water in a sink to whirlpool clockwise down the drain in the Northern Hemisphere, counterclockwise in the Southern Hemisphere. Sailors feel the tug on their fishing vessels, known as the Coriolis effect. Computers make the necessary calculations nowadays, but there was a time when the ability to factor in the Coriolis effect could get a soldier battlefield promotion.

Winds are set into motion by Earth's rotation. Trade winds challenge and help as sailors cross the Equator. They shape their force into tides, and tides are so routine that the Nautical Ephemeris can predict them to the hour and minute for any location on Earth. There are rivers in the ocean, the Gulf Stream, for instance.

Benjamin Franklin was approximately the first scientist to chart the rivers, currents and ocean winds. He made it a matter of record that American ships took two weeks less to cross the Atlantic than British ships took coming the other way. Because of Franklin's work, American captains avoid the drag of currents by sailing around them. Franklin created the term "Gulf Stream." It was too big to be missed. It starts at Miami at five miles an hour, this 50-mile wide stream with a 1,500-foot drag, and it moves about 4 billion tons of water a minute, or about 1,000 times the flow of the Mississippi. At Cape Hatteras, the Gulf Stream heads northeast toward Europe. It challenges the ice waters of the Labrador current off the Grand Banks of Newfoundland.

There are other anomalies. The Sargasso Sea is at least 4 feet higher than surrounding waters. Surrounded by living water, it is a biological desert, so warm nutrients held in escrow below cannot make it to the surface.

There are equally turbulent mixer currents, streams and waves in the South Atlantic, the Indian Ocean and the Pacific, great tsunami towers of water indeed.

After World War II, the U.S. Navy set about the business of studying the ocean floors. Maynard Murray asked for and got samples from every ocean and from many places in each ocean. The samples were remarkably uniform, as though the many hot and cold waters of the world had been powered throughout a gigantic blender.

There were 90 elements in that inventory or blend, all held in perfect suspension. The elements that only ocean water has were there, as was an array of microorganisms.

Nor are ocean waters ever overloaded (except temporarily after a volcanic eruption), as is often the case when human beings and livestock intake minerals beyond homeostatic limits. The ocean has a way of sorting out the excesses.

There are miles and more miles of ocean floor covered with disk-like balls — manganese modules. They are also rich in important minerals the ocean mix finds surplus — copper, nickel, cobalt, zinc, molybdenum, lead, vanadium, titanium. No one really understands the why and wherefore regarding these stores, except to credit nature's designer unless, as is often the case, one chooses to hold that nature designed itself. It may be that submarine vulcanism figures. This sounds reasonable. Certain metabolic oxide molecules colliding with seawater might well find a nucleus the way water vapor uses aerosol for a nucleus. The molecules probably take on their onion-like layers from materials saturated water cannot hold. Manganese molecules put abundant oxygen on the second side of the proverbial equals sign. It has been estimated that 16 million tons form on ocean floors each year. Mined, this sluice box could make pale into insignificance all the land-based manganese mines in the world. When trawlers prowl the oceans of the world, they are always on the hunt for an aquatic Serengeti, the areas of blue water in which sea life congregates, much like exotic animals on the fabled African savannah. Vast schools gravitate toward the upper northern hemisphere. Sometimes great vents in the Earth's mantle spill hot contents that are rich in minerals, as is the case off the Baja Peninsula. Sometimes underwater volcanoes turn several square miles of ocean into boiling soap that becomes food once it cools and submits to water's calming touch. Hot spots off Florida, Hawaii and Australia were full to the brim with turtles, sharks and large ocean fish, tuna — many to staff the endangered list within a few decades. Maynard Murray's scalpel had access to them all. Tuna was perhaps the only market fish he pulled in from these feeding grounds, and they all exhibited virgin health. Fishermen sell many of the rest "mostly by catches." They spawn where nature decrees, reason unknown, and become candidates for extinction at the hands of commercial fishermen, or candidates for preservation at the urging of conservationists. Just the same, there is a demand for hammerhead sharks and shark soup. Even as early as 50 years ago, improved equipment and industrial dedication to harvesting free raw materials, even shark cartilage in health food stores, would see the hammerhead population, all the picture of health when harvested, decline 89 percent between the passing of Maynard Murray and the present. Much the same was true for the blue marlin, Hemingway-sized marlin. As these lines are set down, few of these fish survive big enough to reach maturity, but the survivors do not succumb to disease, as is the case with pond-reared catfish and river-spawned salmon.

Murray saw nets set for yellowfin tuna scoop up dolphins, a consequence of the species schooling together. Necropsy findings were the same: Murray took perfect livers, pancreata, spleens and other internal organs. There are turtles that lay their eggs on the sandy beaches of Costa Rica. If they make it to the sea, they spend years circumventing the globe to return to the hatching grounds to repeat nature's process, all apparently in perfect health. Plankton is usually the food for fish not of the predator class.

Much of what has been set down here is oral history harvested from conversations with the globe-trotting physician himself, from his earliest field-crop assistant and from his public remarks at an Acres U.S.A. conference.

There was more to Maynard Murray's quest than reaching beyond signs and symptoms to first causes. During the 1930s the supermarket came on line, this business scheme later supplemented by the chain store, often a supermarket chain. The byword became "shelf life" — shelf death, actually — of packaged foods, processed foods, foods deficient in vitamins, minerals and enzymes. Clumsy agricultural practices had twice the bins and bushels for an era of hidden hunger. Maynard Murray was convinced that very few of America's acres were capable of producing clean, nutritious, therapeutic food. He and a Chicago nutritionist friend, Harold Simpson, like the term *therapeutic.* It ought to replace *organic,* which is probably the most misunderstood term in the English lexicon.

How to transport health from the ocean to row crop acres was the challenge.

On shore at many ports of call, Maynard Murray was a wave watcher. He was enchanted by the twice-a-day rhythm of tides coming in and going out. There was something of a miracle in how water could be stirred up by waves, rotated, then return to where it was in the first place. A fishing cork can be hoisted high in the air and redeposited within an inch of where it was before the wave came in. There is nothing simple about ocean waves. Some are little more than ripples in a pond. Some achieve the status of whitecaps. Swells and surfs and breaker tides all carry the ocean mix toward land. The moon's tug on the water is the most powerful. It moves the whole ocean. Maynard Murray discovered at least a dozen places where tidal bulges would spill ocean water over shore berms into so-called tidal pools. Here the tropical heat would put evaporation in motion. The resultant sea solids remained, possessed of six mineral keys if worthy of harvest. Maynard Murray did not identify his index system that separated useable sea solids from the dead salts one finds on the Utah Salt Flats, which are also sea solids placed on deposit perhaps 100,000 years ago.

The inventory of underwater volcanoes was modest at the end of World War II. More recent counts have doubled the number, now perhaps 20,000. Underwater volcanoes blow their tops, sending out tsunami waves 40, 50, even 90 feet high. They crash past all barriers and leave behind the mother lode of ocean solids that Maynard Murray figured might save mankind from cancer. At the end of his odyssey, he brought back samples and set about asking nature to reveal itself.

Here is a state-of-the-art readout generally accepted when Dr. Murray began his studies. How the ocean inventory has been enlarged and its nutrients taken up by wheat grass will be shown in the last chapter.

Mean Percent of Elements in Earth Soil

Element	%	Element	%
Oxygen (O)	49.0	Lithium (Li)	0.003
Silicon (Si)	33.0	Gallium (Ga)	.003
Aluminum (Al)	7.0	Copper (Cu)	0.002
Iron (Fe)	3.8	Boron (B)	0.001
Carbon (C)	2.0	Tin (Sn)	0.001
Potassium (K)	1.4	Lead (Pb)	0.001
Calcium (Ca)	1.37	Cobalt (Co)	0.0008
Magnesium (Mg)	0.5	Scandium (Sc)	0.0007
Titanium (Ti)	0.5	Beryllium (Be)	0.0006
Sodium (Na)	0.63	Cesium (Cs)	0.0006
Nitrogen (N)	0.1	Hafnium (Hf)	0.0006
Manganese (Mn)	0.035	Arsenic (As)	0.0006
Sulphur (S)	0.07	Antimony (Sb)	0.0006
Phosphorus (P)	0.065	Thorium (Th)	0.0005
Barium (Ba)	0.05	Iodine (I)	0.0005
Zirconium (Zr)	0.03	Strontium (Sr)	0.0005
Fluorine (F)	0.02	Molybdenum (Mo)	0.0002
Vanadium (V)	0.01	Uranium (U)	0.0001
Chlorine (Cl)	0.01	Germanium (Ge)	0.0001
Chromium (Cr)	0.01	Selenium (Se)	0.00002
Rubidium (Rb)	0.005	Thallium (Tl)	0.00001
Yttrium (Y)	0.005	Silver (Ag)	0.00001
Germanium (Ge)	0.005	Cadmium (Cd)	0.000006
Zinc (Zn)	0.005	Mercury (Hg)	0.000003
Nickel (Ni)	0.004	Radium (Ra)	8×10^{-7}
Lanthanum (La)	0.003		

Compiled by Gerard Judd, Ph.D.

This is the accepted textbook composition of soil elements. It does not necessarily deal with catalysts that perform a function but are not a part of the final product.

Relative Percent of Elements in Earth Soil and in Moon Soil

Element	Earth Soil	Moon Soil	Moon/Earth
Chromium (Cr)	0.01	0.25	25.0
Titanium (Ti)	0.5	4.2	8.4
Scandium (Sc)	0.0007	0.0055	7.86
Nickel (Ni)	0.004	0.025	6.25
Calcium (Ca)	1.4	7.9	5.64
Yttrium (Y)	0.0028	0.013	4.65
Iron (Fe)	3.8	12.4	3.26
Cobalt (Co)	0.0008	0.0018	2.25
Magnesium (Mn)	0.085	0.175	2.0
Aluminum (Al)	7.1	6.9	0.97
Oxygen (O)	49.0	42.0	0.85
Ytterbium (Yb)	0.00027	0.00018	0.66
Silicon (Si)	33.0	20.0	0.61
Sodium (Na)	0.63	0.37	0.59
Vanadium (V)	0.01	0.0042	0.42
Copper (Cu)	0.002	0.0008	0.40
Gallium (Ga)	0.0015	0.0004	0.267
Barium (Ba)	0.05	0.009	0.18
Potassium (K)	1.4	0.1	0.07
Lithium (Li)	0.003	0.00015	0.05
Rubidium (Rb)	0.01	0.0003	0.03
Magnesium (Mg)	0.5	0.0005	0.001
Strontium (Sr)	0.03	0.000015	0.0005

Partial List of Elements
(Parts Per Million) in Ocean Water

Element	ppm	Element	ppm
Oxygen (O)	857,000	Vanadium (V)	0.002
Hydrogen (H)	108,000	Titanium (Ti)	0.002
Chlorine (Cl)	19,000	Cesium (Cs)	0.0005
Sodium (Na)	10,500	Cerium (Ce)	0.0004
Magnesium (Mg)	1,350	Antimony (Sb)	0.00033
Sulfur (S)	985	Silver (Ag)	0.00033
Calcium (Ca)	400	Tellurium (Te)	0.0003
Potassium (K)	380	Cobalt (Co)	0.00027
Bromine (Br)	85	Gallium (Ga)	0.00003
Carbon (C)	28	Neon (Ne)	0.00014
Strontium (Sr)	8.1	Cadmium (Cd)	0.00011
Boron (B)	4.6	Tungsten (W)	0.001
Silicon (Si)	3	Germanium (Ge)	0.00007
Fluorine (F)	1.3	Selenium (Se)	0.00009
Argon (Ar)	0.8	Xenon (Xe)	0.000052
Nitrogen (N)	0.5	Chromium (Cr)	0.00006
Lithium (Li)	0.18	Thorium (Th)	0.00005
Rubidium (Rb)	0.12	Mercury (Hg)	0.00003
Phosphorus (P)	0.07	Lead (Pb)	0.00003
Iodine (I)	0.06	Beryllium (Be)	0.000017
Barium (Ba)	0.03	Gold (Au)	0.000011
Indium (In)	0.02	Titanium (Ti)	0.00001
Aluminum (Al)	0.01	Tantalum (Ta)	0.0000025
Zinc (Zn)	0.01	Zirconium (Zr)	0.000022
Iron (Fe)	0.01	Hafnium (Hf)	0.000008
Molybdenum (Mo)	0.01	Lanthanum (La)	0.000012
Nickel (Ni)	0.0054	Helium (He)	0.0000069
Arsenic (As)	0.003	Scandium (Sc)	0.000004
Copper (Cu)	0.003	Beryllium (Be)	0.0000006
Uranium (U)	0.003	Protactinium (Pa)	2×10^{-9}
Tin (Sn)	0.003	Radium (Ra)	6×10^{-11}
Krypton (Kr)	0.0025	Radon (Rn)	6×10^{-16}
Manganese (Mn)	0.002		

3

Ocean Traces, Ocean Solids

"I think of the millions we are spending on cancer research and on arteriosclerosis, and I think we ought to be investigating this correlation between sea life and good health. There is a connection, I know it." So spoke Dr. Maynard Murray to associates, business forums, investors in the sea life connection, and not least, to Ed Heine, a polio victim since his teens and now an active farmer in spite of his infirmity.

Count them — there are 90 nonradioactive or trace elements in perfect solution naturally. These elements in varying degrees are necessary to all life forms. In soils, an absence or imbalance of trace nutrients interdicts the development of hormone and enzyme systems, thus incubating bacterial, fungal and insect attack. The river of blood that sustains human beings is a copy of seawater. Why, then, are the same elements not also a requirement for cancer-free living? To ask the question was to suggest an answer.

The answer arrived as a bag of ocean solids harvested from an evaporated pool on the Baja Peninsula. The contents of that bag were as solid as a block of concrete. "It was the first of many problems that went with my research for the next 30 years."

"In 1953 I was a student at Elmhurst College in Elmhurst, Illinois, majoring in history." This is Ed Heine speaking, his thoughts trailing back to when he first met Maynard Murray.

A speech teacher named Ben Jacques was doing preliminary plan research for Murray in a greenhouse owned by the City of Elmhurst. After a few weeks of simple discussion about using ocean salts as a balanced nutrient to aid plant life, Jacques enlisted Ed Heine to try some on his garden and possibly on a small row crop plot. It was the yes that triggered events to follow, one of which was the arrival of that concrete-hard bag of ocean solids. As a handicapped student who had suffered from polio, Ed Heine didn't feel much like swinging a sledgehammer. His forte was using his head.

Always active in the Future Farmers of America – he won an Illinois state speaking contest and served as a state officer in 1947 – Heine received the American Farmer Award in 1948. He understood the parameters of controlled research. His poultry studies had been successful, and he had comfortable dealings with credentialed academia as well as a highly respected physician, a 100-pound burlap sack of ocean solids notwithstanding.

For field application, the ocean solids had to be turned into fine powder. Water might have helped loosen the rock, but the experiment's goal was powder, not soup. Jacques tried to grind it down, but the attempt merely wrecked the 3-inch drive shaft of a commercial hammer mill on an Ohio farm. Thus the challenge passed to Heine. Armed with a three-pound hammer and a small wooden box coffee grinder with a small sliding drawer, "I set about sledging that sack of solid rock into fine powder. It took time to chip chunks off the salt-containing ocean solids, the only constant being atmosphere humidity. Several hours later, his progress suggested the coffee grinder was less than efficient or economical.

Next up was a 10-inch burr mill. The 3-inch drive shaft explosion was still a haunting memory, but if at first you don't succeed, try, try again before you quit! An old 3-inch belt drive instead of the usual 6-inch belt, he speculated, would slip before either the drive shaft or burrs broke. It worked. Fine powder rained down into the receiving container. In about four minutes,

97 pounds of fine salt powder was restored to its pristine ocean splendor.

Jacques told Heine to apply one pound to an area of approximately 15 square feet, then incorporate it into the soil at about one ounce per square foot. These early experiments were at least as painstaking as anything pursued at an official experimental station. The garden came first. Tape-measured sections of rows of each vegetable variety were treated, literally spoon fed! Next came the field corn, each plot 1/100th of an acre. Using a plot high spot near the end rows of a 40-acre field easily accessible to a person using crutches and braces, 30 pounds of material were applied to two 40-inch wide rows of corn that was already 2 to 3 inches tall. Along each row, the ocean solids were released into the soil. Metal fence posts marked the treated areas. Untreated controls remained in place during the experiment.

A small pasture area was also treated; on it was a fruit tree. This the farmer used as an aside, an expedient way to dispose of the powder not needed for the basic test.

Dr. Murray took a very personal interest in the ongoing response, "and that's the first time I met him," Ed Heine said. Murray was pleased with the fineness of the material. He admitted that there was deficit to making ocean solids into fertilizers. The salts were from sun-drenched tide pools of routinely captured ocean water, which evaporates and leaves ocean solids behind. These are quite unlike the salt flats of Utah. He admitted that previous attempts at setting up small-scale projects had faced the same impediment.

Later on, salt chunks imported for experimental purposes were hammered into submission on a concrete pad with a roto-vator, and field spreading was done by a regular manure spreader, the best available working tool. This system was aborted because applications could not be controlled.

Heine's experiments were a failure. Jacques had deposited the application at toxic levels. The correct rate of application should have been one pound per 20 square feet, not 15 pounds for 20 square feet.

Maynard Murray did not come out very much that year, Heine recalls. When he did, it was to scrutinize the corn test plot.

The plants weren't painted green, as is the case with heavy nitrogen. Some stalks were distressed. Some died. At most the ears were short nubbins, in fact. Many stalks had no ears at all. The entire plot of tape-measured rows produced only a few gunnysacks of corn.

In terms of bins and bushels of cosmetically beautiful corn, the season's trial was a failure. And yet there was a collateral result that reminded him about the real purpose of corn, which was to pick up nutrients, trace nutrients, nutrients denied hybrid corn versus open-pollinated corn. During the Depression years, tests were done by the old Armour Laboratories in Chicago, on OP corn versus hybrids. Those results now came to mind.

A feeding test on the Heine farm was to present a few ears of the test corn and a few ears of the regular corn for a feed mix. Each ear was broken in half. The many halves were then mixed together and dumped into a fence line feed bunk. Cows in the pen nuzzled the ears. They always selected the test-plot nubbins, then finally consumed the rest. Had they been able to eat their fill of nubbins, the rest would have remained in the bunk uneaten.

It was a case of back to the drawing board after this first season. Obviously, the elements were there, which is seldom the case when land has been insulted with salt fertilizers and toxic rescue chemistry. Somehow the variables had to be brought under control. The corn was a standard hybrid. What could OP corn do? The land was not bio-correct; that is, it was under treatment with N, P and K. Rainfall is always a factor. Finally, here you have corn bred to withstand starvation—the reason for being of hybrids—and suddenly a cafeteria of nutrients becomes available.

Ninety-two nutrients? That is a bunch. Plants are very selective in what they uptake. Often they want to uptake nutrients not even present, causing rusts, viruses, and diseases. Weeds are specialists. Usually they survive and proliferate by whole-hogging a nutrient, burdock uptaking iron, for instance. It takes the right combination to feed a row crop plant or pasture grasses and forbs, and Maynard Murray felt certain the ocean had achieved that mix.

How to transport, say, molybdenum from the ocean into the corn plant, then into the animal, and into the human diet? That was a sample question. Researches identified molybdenum as essential, and yet it is almost totally missing in the soil. An aneurysm that stands ready to bring down a human being reveals a copper shortage, copper meaning ocean copper. Micronutrients key the enzymes, their absence permitting allergies, hay fever, asthma, high blood pressure and loss of mental acuity.

Copper has gone unrecognized. It backbones *Soil, Grass and Cancer,* the André Voisin book on missing pasture nutrients that lead to cancer. Copper is antagonistic to parasites and intestinal-tract worms. It is estimated that over 95 percent of all Americans have microscopic parasites and are unaware of this fact. The cancer risk haunts medicine and frightens the population; one out of four can expect to endure that select tormentor and killer at some point in their lives. Always a gem of a nutrient in the ocean mix, the amount of copper transported into the human bloodstream is elevated in an ocean water diet.

Dmitri I. Mendeleyev, the Russian chemist, first constructed his table of elements more than 100 years ago. His insight into the natural order was so profound he supplied spaces where he believed an element would be discovered to comply with the rhyme and reason for the supreme plan. That table still stands. The blank spots have been filled in, and the Periodic Chart of Elements has provided a simple and beautiful picture of the order of our universe. It has done more. It has opened chemistry and physics as never before and made it possible for ordinary people to understand the structure of the atom. It has provided facts that agnostics could not ignore. Each element's entry is an inventory of information, starting with hydrogen, the lightest element. The atomic weight on the chart is almost always expressed as a single-figure average when isotopes are involved. Isotopes are atom's brothers, so to speak, atoms of the same element that differ in weight. Hydrogen has three isotopes. Protium hydrogen, the highest form, has one proton and one orbiting electron. Thus a slightly heavier form is deuterium, also called heavy hydrogen, and the heaviest form is called tritium. The

Average Percent of Elements
in Earth's Crust (10-Mile Shell)

Oxygen (O)	46.6	Tin (Sn)	0.004
Silicon (Si)	27.7	Yttrium (Y)	0.0028
Aluminum (Al)	8.13	Neodymium (Nd)	0.0024
Iron (Fe)	5.0	Niobium (Nb)	0.0024
Calcium (Ca)	3.63	Cobalt (Co)	0.0023
Sodium (Na)	2.83	Lanthanum (La)	0.0018
Potassium (K)	2.59	Lead (Pb)	0.0016
Magnesium (Mg)	2.09	Gallium (Ga)	0.0015
Titanium (Ti)	0.44	Molybdenum (Mo)	0.0015
Hydrogen (H)	0.14	Thorium (Th)	0.0012
Phosphorus (P)	0.118	Germanium (Ge)	0.0007
Manganese (Mn)	0.1	Samarium (Sm)	0.00065
Sulfur (S)	0.052	Gadolinium (Gd)	0.00064
Carbon (C)	0.032	Beryllium (Be)	0.0006
Chlorine (Cl)	0.031	Praseodymium (Pr)	0.00055
Rubidium (Rb)	0.031	Scandium (Sc)	0.0005
Fluorine (F)	0.03	Arsenic (As)	0.0005
Silicon (Si)	0.03	Hafnium (Hf)	0.00045
Barium (Ba)	0.025	Dysprosium (Dy)	0.00045
Zirconium (Zr)	0.022	Uranium (U)	0.0004
Chromium (Cr)	0.02	Boron (B)	0.0003
Vanadium (V)	0.015	Ytterbium (Yb)	0.00027
Actinium (Ac)	0.013	Erbium (Er)	0.00025
Nickel (Ni)	0.008	Tantalum (Ta)	0.00021
Copper (Cu)	0.007	Bromine (Br)	0.00016
Tungsten (W)	0.0069	Holmium (Ho)	0.00012
Lithium (Li)	0.0065	Europium (Eu)	0.00011
Nitrogen (N)	0.0046	Antimony (Sb)	0.0001
Cerium (Ce)	0.0046	Terbium (Tb)	0.00009

Lutetium (Lu)	0.00008	Palladium (Pd)	0.0000017
Titanium (Ti)	0.00006	Platinum (Pt)	0.0000005
Mercury (Hg)	0.00005	Gold (Au)	0.0000005
Iodine (I)	0.00003	Helium (He)	0.0000003
Beryllium (Be)	0.00002	Tellurium (Te)	0.0000002
Thulium (Tm)	0.00002	Rhodium (Rh)	0.0000001
Cadmium (Cd)	0.00001	Rhenium (Re)	0.0000001
Silver (Ag)	0.00001	Iridium (Ir)	0.0000001
Indium (In)	0.00001	Osmium (Os)	0.0000001
Selenium (Se)	0.000009	Ruthenium (Ru)	0.0000001
Argon (Ar)	0.000004		

average of these several weights is 1.00797. Each of the elements of life and death has an observation symbol. Some appear on fertilizer bags, in farm literature, even as slang in daily conversation.

A chart of elements, usually all those available in the ocean, exhibits numbers in brackets. This means the numbers represent an isotope of the element, usually the one with the longest half-life. These numbers have been studied by the National Bureau of Standards. Published numbers are standard according to the International Union of Pure and Applied Chemistry. All the elements generally accepted as necessary for life are listed as the first 53 or 90 natural elements, all of them ocean elements.

Of these, all except one fall in order among the first 42. All except two are listed among the first 34.

There is also a natural order of abundance according to the atomic weight and number. The heaviest elements are the rarest. Elements with even atomic numbers are more abundant than those with odd numbers. We do not know why, nor can we even guess. The table itself is a veritable encyclopedia. There are series with missing electrons as the eye moves from titanium to zinc. Orbits change an electron one at a time. These limitations take place in a material order, moving across the table. There is also a vertical order to the table, weight increasing as each element is listed under the one above. There are groups that figure in biology and signal the entrance and exit of disease. Henry A. Schroeder, the world's foremost authority on trace nutrients at the time of his death, left his research in a book, *The Trace Elements and Man*. He noted that a heavier metal can displace a lighter one in the same group in biological tissue and alter the reaction of the lighter one. He went on to say that tissues with an affinity for a certain element have an affinity for all other elements of the same group. Some elements are bone seekers. Some are thyroid seekers. All elements in two groups are liver and kidney seekers. In terms of plant life, it is too early to say which of the elements are essential. Perhaps they are all essential, even though college texts and agronomy manuals are fond of listing 14 or 16 or 18, sometimes more. If you find a periodic chart on a doctor's wall, you will note that plus symbols designate the footnotes, and the footnotes say the atomic weights are reliable to plus or minus three in the last place. Other weights are reliable to plus or minus one in the last place. These are isotopes. Isotopes are atoms with the same number of protons but different numbers of neutrons. Mass can be identified by the spectrometer, a relatively modern instrument that delivers an ocean water readout in less time than it takes to tell about it. It was the development of this instrument that enabled Maynard Murray to serve up answers to questions about whether oxygen was released from plant life or from water or from carbon dioxide.

Dr. Lewis Thomas, writing in *The Lives of a Cell,* makes quite a point of the proposition that the human system is not alone as an operating mechanism. "We are enslaved, rented, occupied," writes Thomas. "The very interior of our cells are homesteaded by mitochondria. Once separate creatures, mitochondria may or may not have been early pre-ancestors of eukayrotes cells and stayed on for a few billion years." These little fellows are sufficient servers. They have their own DNA and RNA and replicate in their own way. Here the rhizobial bacteria govern the rate of reproduction. They are symbiants. Except for them, we couldn't drum a finger, much less think a thought.

Plant life is not plant life as such either. It too is rented out and occupied. Little one-cell creatures are everywhere; they are small and invisible to the naked eye. Plant and human life depend on them, suffer disease because of them, live and die according to how life and death are balanced.

Chloroplasts work with the sweep of the sun to manufacture the oxygen we breathe. We will encounter oxygen again as we unveil the leap to enzymes. For now it is enough to remember the periodic order of trace nutrients, the key to every enzyme ever discovered, for enzymes operate life's work, its health profile, shelling out debilitation and death according to how they are preserved and managed.

Early in the 20th century, researcher and physician George H. Earp-Thomas proved that plants allow the transport of minerals through the root hairs when they are ions in size. This was confirmed in the University of Missouri laboratory of William Albrecht and his graduate students. Micron-sized nutrients cannot make the trip through a plant's roots and therefore cannot indirectly service the nutritional needs of bovines. It is the supreme function of microorganisms to ready the earth minerals for plant uptake, enzyme construction and human consumption. The cobalt Earp-Thomas found missing in almost all soils left its imprint in animal health as brucellosis, and in humans as undulant fever. The law of reasonability decreed that the missing cobalt must be supplied. Instead, the Department of Agriculture duly set up a brucellosis eradication program that required annihilation of infected herds in order to make communities and areas brucellosis-free.

A shortage of molybdenum makes it difficult to expel waste hydrogen from the belly. Further, dealing with so-called mineral-box ailments by supplying calcium and phosphorus can complex and confuse the trace mineral molybdenum even if it is in the soil, which usually it is not, installing nutritional frustration in the human diet.

The components of supplementation are often so numerous that micronutrients cannot negotiate their way around antagonisms of the pecking order.

Position 27 on the Mendeleyev chart is cobalt. Position 42 is occupied by molybdenum. Copper is stationed at 29.

These are only examples of the nutrient level. Count them, there are 90, all of them available in ocean water, few distributed commercially in useable form for row crop acres, free ranges or manmade turf pastures.

The Mendeleyev chart is more than a blueprint of nature. It is a nutritional chart of the ocean, the very ocean Maynard Murray believed was embodied in the ocean minerals deposited in shallow tidal pools and dried by the sun.

4

Stepping into the Light

Dr. Maynard Murray did not wait for field trials to confirm his insight. The absence of malignancies in ocean creatures was more than a puzzlement; it was the debut of a new passion in Maynard Murray's soul. Concurrent with the few greenhouse tools provided by Mr. Jacques, Murray wondered about whether hydroponics with ocean solids dissolved in water would not better service plants than worn-out soil. Why not grow vegetables in fertilized water rather than soils? In his basement he set up trays and lights. He soon found what all hydroponics growers already knew, that hydroponics plants grew better under full-spectrum lights. Tomato plants thrived in test water and produced fruit for 14 to 16 weeks. Indoors or not, it made no difference as long as ocean nutrients were forthcoming. Plants could be crowded together without diminishing yields. The numbers stacked up in his notebook. A 10 to 50 percent increase seemed normal. For the next three decades, the good doctor had fresh produce the year-round.

"I think of the millions we're spending on cancer research," he told a dozen or so investor friends. The same was true for arteriosclerosis. "I think we ought to be investigating this connection between sea life and good health. There is a connection, I know it."

It was the presence of 90 elements in ocean's balanced mix that had taken command of his voice and emotions, and he wouldn't let it go, not for a medical practice that could have made him one of the top 5 percent of income earners in the nation. He could not dismiss the idea that all these perfectly sized and balanced nutrients were necessary to all life forms. Soil and water tests revealed that modern farming had become a form of mining, doubly so because the tax people did not permit the farmer to take a depletion allowance, as was the case with extractive industries. By the early 1950s academic advisers had instructed growers to balance calcium, magnesium, sodium and potassium to achieve a 6.8 pH, to use a salt form of phosphorus, to more or less ignore the natural nitrogen and carbon cycles and to expect to short-circuit life's electrical system. This conventional ignorance almost entirely ignored the traces, even though the same intellectual advisers agreed that the nutrient in the least supply held in escrow plant, animal and human health. The absence of full-spectrum nutrition invited the use of toxic rescue chemistry as bacterial, fungal and insect attack asserted themselves. Their role, after all, was to remove crops judged unfit to live by nature. Rescuing such crops with toxic sprays took on the aura of insanity, and Maynard Murray knew it.

Minerals were the key to enzymes. Everyone talked about enzymes, but like Mark Twain's weather, few people did anything about them. Their discovery and each function was proceeding at a pace even slower than discovery of the role each mineral played.

All enzymes have trace mineral keys, keys that can be made radiomimetic by radar ranges, irradiation and the obscene presence of farm chemicals that disturb the soil environment. Even now, scientists do not know or have not released results on what nuclear radiation has done to Earth's soils. Such readouts could be available because Missouri's William A. Albrecht sequestered soil samples before these mushroom clouds rose over Hiroshima and Nagasaki. Feeding his basement plants sea solids rather than N, P and K separated Maynard Murray's homegrown experiments from the greenhouse hydroponics push that was sweeping the nation.

Over 50 years have come and gone since these ill-fated field trials on the Ed Heine farm. The many years that were to follow showed the somewhat disabled farmer a glimpse of true scientific dedication far beyond what medicine had to offer. "Little did I realize that for 29 years I would see and absorb just a tiny portion of that vision," farmer Ed Heine said. Years and results separated each latest finding from the stone hammer and coffee grinder days, and conclusions flowed from the facts.

After a year of dabbling with ocean solids on small plots, the experimental design took form. Trips to the Heine farm cemented relationships, and Maynard Murray concluded that the right time and the right place had come into focus. Ed Heine revealed the passage of events.

"Dr. Murray approached my father, Ray Heine, about the potential of having a larger farm area for experimental and controlled field plots." He would provide the funding for a special sea solids spreader. It was essentially a two-wheel, 10-12 foot dry fertilizer spreader that was commercially available. A several thousand-bushel metal grain bin and a rotovator and a corncrib were also ordered in to accommodate the trials. The Heine family was to provide the acreage for control and test plots. A rail car would deliver the ocean solids from Mexico. The solids also had to be ground into fine dust with the burr mill that had proved successful in the past. A computed 2,200 pounds per acre plot was to be the fertilized bed. The land would be treated and the ocean solids disced or tilled into the soil before planting. The same seed variety was to be used on both the control and experimental plots. Oats, corn and soybeans were the crops planted.

When spring arrived, the fertilizer spreader and grain bin were in place and the sun stood ready to deliver its solar revenue. Maynard Murray had elicited the financial help of 11 "investors," actually contributors who each added a token amount to the cause of the experiment. All were executives or owners of large businesses.

In the Baja Peninsula, a rail car of select ocean solids was loaded and shipped. It crossed the border at Calexico, California, rolled east to New Orleans and was promptly lost somewhere in the railway yard. Days and weeks passed. Ed's father, Ray Heine, informed Murray that the fields were ready for oats and that planting had to proceed even if no rail car arrived. This was early in the morning at 10:00 a.m. Maynard Murray contacted one of the 11 investors, a railroad executive. "What happened?" was more than a rhetorical question. Without that rail car, another year would go up in smoke. The railroad executive found the car at 2:00 p.m., aging in New Orleans. It had been sitting there for three weeks, buried into oblivion by inept business bureaucrats. At 11:00 the next morning a passenger streamliner stopped at Plato Center, Illinois, and backed into a rail siding to uncouple a boxcar. It was a sight the station agent would not soon forget — a sleek streamliner, the *City of New Orleans,* unloading a time-worn boxcar. The station agent commented, "Someone with real authority had to authorize this to get the car transported this quickly."

Once the car was unloaded and the ocean solids were ground for the experimental oat field, the dry-material spreader had to be calibrated. A collection pan was placed under the outlets of the spreader. Trial and error finally determined the correct calibration. Everything worked well the first day, when weather and timing were perfect. By the next morning, a front had moved in. Clouds darkened the sky, and humidity gave the ocean solids a signal to cake and clog access to the machine's agitators. A platform had to be built for workmen to break the cakes as they formed so the material would continue to fall into the agitators. Also, the calibration had to be increased. Thus surfaced a real deficit. It took very little humidity to interdict free fall of the dry product. But the soil ate up the fine product with gusto, if such a term can be used. Particle sized governed. The smaller the grind, the slower the absorption — not a problem when planting could be delayed.

Maynard Murray made the 35-mile run to the Heine farm on a regular basis during the trials, observing germination, growth, maturity and harvest.

Corn and soybeans followed oats in the planting sequence.

The time line for reports and evidence has become blurred by the years. Ted Whitmer, a farmer and a consultant from Glendive, Montana, picked up on the ocean solids idea and asked the Heine family for enough of the materials for experimentation on his own. Whitmer was a lone hetman. He packaged his wheat and sold it at retail while all the other wheat farmers were allowing their capitalization to slip away. He followed farm developments and reported back, and ocean solids were high on the agenda. His own oral testimony to this writer told of identifying collection sites for ocean solids, and pasture trails showed promise. The mix and the rainfall were vital, as was the tolerance of plant species.

A nephew stationed in Algeria reported a failure of irrigation pumps. Ocean water was used in place of deep well water only once during the pumpkin-growing season. A bumper crop sans disease became the observed result. During three decades of association with Maynard Murray, the reports and fallout stacked up like cordwood. Ed was off to a seminary a part of the time while his father carried on. Part of a caseload of ocean solids was shipped to New England, where a 17-year-old artificial insemination bull no longer provided semen capable of settling cows. In two months on grain grown with ocean solids, that bull could settle a cow that finally delivered a healthy calf again. Publication of the Tom Valentine book was at its peak when this report became current coin, hence its omission in *Sea Energy Agriculture*. A second book, this one by Maynard Murray himself, was at the publisher when he died in 1982. His family stopped publication and the manuscript was lost.

Not lost were these two full-hour conversations Maynard Murray paused to have with Ed and Ray Heine when he came out to visit his trials on Sundays. He not only surveyed the land, he tutored. He expanded the knowledge of those who had the will to listen.

For the Illinois experiments it was decided that 2,200 pounds of ocean solids per acre was the optimum. The matter of dilution had its starting point. Cypress trees in the Florida Gulf area grow in 100 percent seawater. So do the mangrove swamp trees that made parts of the peninsula so hostile to Ponce de Leon as he scoured the countryside for a fountain of youth, never realizing that the fountain of youth was with the very ocean on which he sailed. Vegetables have their own demand, as does wheat grass, the pasture with its several dozen species, fruit trees and plants grown away from the ocean in terms of space and time. At 3,000 pounds per acre, ocean solids proved mildly toxic for most crops, hence nubbin corn, for instance. Also, there were other contributing factors: the soil and its cation exchange capacity, the blessing of rain. In sandy soil, ocean solids dissipated with the water.

Results from the several experiments conducted by Ed Heine and Maynard Murray have been made a matter of record on pages 46 through 48 in *Sea Energy Agriculture,* and are included here:

1. Corn experiments:
   a. Sea solids were in no way detrimental to the growth of the corn.
   b. Uniformity of growth:
      - experimental corn – substantially free of nubbins, uniformly high stalks;
      - control corn – usual distribution of nubbins, normal variation in size of stalk.
   c. Yields:
      - experimental ears – 1.5 inches longer on the average than control ears, 3/8 inch larger in diameter on the average than control ears.
   d. The experimental plot yielded four more bushels per acre than the control plot.

2. In a garden experiment, complete sea solids were applied to a 10-by-20-foot plot and worked into the soil before planting radishes, beans, peas, carrots and lettuce. The same plantings were made in a control plot not fertilized with sea solids. All vegetables grown in the experimental plot had a superior taste to those grown in the control plot, and the leaf lettuce of the experimental area permitted four cuttings compared with two cuttings of control lettuce.

3. During the next growing season, which followed the preliminary experiments outlined above, the following larger-scale field experiments were conducted.

a. Oats:

April 19-24 – Sea solids ground in burr mill to a very fine texture were applied to soil using an International Harvester Ten-Foot Fertilizer Spreader. The 2,200 pounds per acre of sea solids were spread over 10 acres of a 19-acre field, leaving a nine-acre portion of the field untreated. The sea solids were worked into the top 4-7 inches of soil using a 12-foot field cultivator, and Bonda oats were broadcast and disked in the complete 19-acre field. Heavy rain fell intermittently through June 4th.

May 3 – Observed oats were coming up; control oats appeared to be taller than experimental.

May 7 – Control oats were 1 to 1.5 inches taller than experimental.

June 7 – Oats in both plots approximately 9 inches high.

June 10 – Color difference observed, and the exact line where fertilization stopped was apparent through the center of the field. Experimental oats had a much darker green color. Rabbits and grasshoppers were observed to exhibit a marked preference for oats in experimental plot.

June 13 – Cows being driven down the road preferred grass at the edge of experimental plot.

June 14 – Color difference of oats is more pronounced.

June 18 – Oats headed out with experimental oats more advanced.

July 21 – Oats in experimental plot are ready for cutting.

July 24 – Oats in both plots were cut; experimental oats were found to have less rust.

Yield:

Control Plot: 38 bushels per acre.

Experimental Plot: 45 bushels per acre.

b. Corn:

May 25-30 – Manure was applied to 30 acres of a 40-acre field on May 25. In May 2,200 pounds of sea solids per acre were applied in the manner outlined above to a 10-acre plot, retaining the remaining 30 acres as the control plot.

June 8-9 – Pioneer corn planted in entire field, along with 50 to 80 pounds per acre of a nitrogenous fertilizer material (commercial 2-12-12 fertilizer).

June 14 – Corn showed above the ground and no apparent difference between experimental and control was noted.

July 22 – Corn was observed to be tasseling.

August 1 – Tasseling of control corn was further advanced than the experimental.

August 23 – Corn in both plots was observed to be the same height and color. Each hill of corn on 4.9-acre portions of experimental and control plots was inspected for galls (smut). Control corn had 384 percent more observable galls than experimental corn.

Yield:

Control Plot: 75 bushels per acre.

Experimental Plot: 88 bushels per acre.

4. During the next season, 360 day-old New Hampshire chickens were obtained for feeding experiments using soybeans, oats and corn grown on the sea solids fertilized soil during the previous season. The control group of 153 chicks was fed commercial concentrate plus a feed mixture of two parts corn and one part oats grown on control plots. The 153 experimental chicks were fed the same mixture as the control group, with the exception that the corn and oats used were grown on the experimental plots, and thus were fertilized with sea solids. The following results were obtained.

New Hampshire Chicken Feeding Experiment

Roosters	**Control Group**	**Experimental Group**
(average weight in ounces)		
At 4 months	42	60
At 6 months	10	128
At 2 years	135	152

Hens	**Control Group**	**Experimental Group**
(average weight in ounces)		
At 6 months	80	104
At 2 years	96	114
Time of Laying	5 months, 3 weeks	5 months
Eggs (weight/dozen)	19-23	23
Time of Laying	After 7 months	
Eggs	241	28

Entire Group	**Control Group**	**Experimental Group**
Average feed consumed per lb. of weight gain (lbs.)		
	3.0	1.89
Size	Varied	Uniform
Disease		
Worms	Yes	No
Nervous Condition	Yes	No
Leg Disjointing	Yes	No
Mortality	3	0

Field Crop Ash Weight (% Solids)

Sample	Control	Experimental	Increase (%)
Onions, bulb	13.6	14.2	4.4
Oats	87.7	87.8	0.1
Sweet potatoes	28.8	31.2	8.3
Tomatoes	4.8	5.7	18.7
Soybeans	73.9	84.7	14.6
Corn	73.1	74.4	1.7

5. One sow and six pigs raised on corn and oats grown on land fertilized with complete sea solids were unusually uniform in size, showed no tendency to "root" and were easily contained in a small fenced area. When they reached approximately 180 pounds, they were taken off this feed and given control corn and oats. They immediately began extensive rooting and, by the end of the third day, they were extremely nervous and broke out of the pen on two occasions. On the fourth day they were put back on sea solids grown feed and were calm by evening. Thereafter, they were easily contained in the pen and again showed very little rooting tendency.

The table above shows the ash weight determination of field oats, corn and soybeans, tomatoes, sweet potatoes and onions that were raised on a garden plot at Elmhurst College, Elmhurst, Illinois, with the equivalence of 2,200 pounds of sea solids per acre on half the garden. The other half was fertilized normally.

Field tests were conducted in Wisconsin, Illinois, Ohio, Pennsylvania, Massachusetts and Florida. Always, production was greatest on plots and fields treated with seawater or sea solids, this compared to controls. The same general pattern was observed when animals were fed on feed of all types grown in solids from the ocean.

A factor emphasized herein is taste. All the garden vegetables exhibited superior taste. Onions, potatoes, tomatoes, sweet potatoes, apples and peaches were "outstanding," according to narrative summaries generated by always ongoing research. "Onions could be eaten like apples," one arresting summary said.

Not lost was the enlargement of knowledge concerning trace nutrients, their clandestine role in the Creator's crop plan and their input on human mental acuity. Conventional agriculture has its norms, to be sure. Always, ocean solids modified these norms, as the paragraphs that follow will illustrate.

These results enlarged settled science that was already a part of farming's best kept secrets. In any soil system, excessive magnesium will cause potassium, phosphorus and nitrogen deficiency. Excess potassium, sodium and magnesium will cause calcium deficiency. Excessive calcium will cause magnesium deficiency, phosphorus and trace element deficiency. Excessive magnesium will cause effects similar to magnesium deficiency. Excessive boron will cause potassium and magnesium deficiency. Excess sodium and chorine will cause potassium.

It can be seen that excesses of several stripes can upset cation and anion balances, often the result of careless and indifferent use of N, P and K fertilization.

Phosphorus and sulfites are also of major importance. Still, it would cost a farmer too much money to run a comprehensive test on manganese, boron, zinc, copper, iron and so on. These can be applied safely via the agency of sea solids without testing when a farmer has enough magnesium, potassium, phosphorus and calcium all in the right balance, that is, 65 to 75 percent of the soil colloids being loaded with calcium, 15 percent magnesium, 5 percent sodium. The trace minerals supplied by ocean solids help finish out the 5 percent.

There are deposits in the Mississippi Valley that supply trace minerals in a well-regulated form. Nutrients in the correct balance give a great assist to problems such as compaction and shortage of humans, but do not outpace ocean minerals in the work of supplying the great range of trace minerals required for enzyme construction.

When major elements are in relative balance, it still proves much cheaper to supply trace nutrients in a broad-spectrum way with ocean solids rather than to deal with zinc, iron, manganese, etc. A starter supply of nutrients around the seeds should guarantee the nutritional needs of essential elements throughout the life of the plant. The harvest of 10 to 20 percent more seed from the use of even one sea solids treatment is now a matter of record. Competent farmers have come to obtain such yields on acres too numerous to be considered a test plot.

Often, poor soil cannot be balanced in a single year. It is precisely such a condition that ratified the use of ocean solids. As biologically correct fertilizer, ocean solids can make up for the imbalance and limitations and produce better crops while soil conditions are repaired. It does not seem possible for a hostile environment to complex the nutrient load contained in ocean solids. These nutrients are always available to start the germinating plant with more vigor, vitality and more complete natural absorption capacity. The cell wall structure can be more efficient for the intake of nutrition.

It will undoubtedly be a requirement for farmers to know as much about the soil as they do about the science of agriculture. He or she will have to let the soil tell what is right as well as what is wrong, the way an old-time physician looked to signs and symptoms rather than laboratory results.

The givens are clear enough. Healthy plants cannot grow in a sterile soil, nor can they grow in a soil not balanced and buffered to synthesize balance. That is why American farmers grow a surplus of sick plants and why they rely on toxic rescue chemistry to keep insects away.

No less than six trace elements comprise the one trace element group. They are iron, manganese, zinc, copper, boron and molybdenum. Without exception, plants require these elements for nutritional support, and their absence or deficiency in sea solids condemns that kind of supplement to a deficit. Without exception they are indispensable for some psysiological function, such as the buildup of enzymes. If the trace element is not present, the specific function is not produced.

There are the so-called secondary trace nutrients, such as aluminum, chlorine, sodium, iodine, boron and silicon. These closely influence growth in some plants, but not in all, meandering from class to class. This statement must be hedged, because the above is true as far as our primitive knowledge understands it today.

There are agriculturists, Maynard Murray included, who figure all trace nutrients have a role in plant protection and health, and only our veil of ignorance keeps us from full knowledge of each role. Certainly each decade sees nature revealing herself a bit more. She has been doing so for a long time.

The Dimtri I. Mendeleyev Periodic Table of Elements was one of several discovered at approximately the same time, each setting up the periodic order of the elements that make up the universe. In human medicine, Bertrand's Law and Weinberg's Principle are used to explain why some trace elements are necessary and yet become toxic. The explanation is contained in the term *homeostasis.* Where trace elements are entirely absent, there is no growth. As concentration increases, the deficiency area is diminished. When the level of some trace minerals achieves a plateau, function and nutrient availability are at a peak. The width of the plateau varies with the life form involved. In mammals it is quite wide, because natural trace nutrients are under homeostatic control. The human body has a fantastically exact system for maintaining balance of its essential composition; plants are not as fortunate. For this reason the plateau for plants is narrower. Beyond the plateau needed for proper growth, the same nutrients that might have been characterized as sufficient at one time are now toxic, even lethal. Indeed, trace minerals for plants are unlike those under homeostatic control for human beings. Chlorinated hydrocarbons and many manmade toxic genetic chemicals are not under control and may not be for hundreds of centuries of human evolution.

With fossil-fuel firms pouring manmade molecules into the environmental market in excess of 1,000 new compounds each year, man is not likely to adjust to this pollution factor. Conversely, the organic elemental content of human food — carbon, hydrogen and oxygen — are under poor homeostatic control. This is why these elements build up and deposit themselves in fat. When people eat more than they burn, they might as well add the new poundage directly to their hips.

Trace nutrients blend into the farmer's art, because almost all can be used to detect hunger signs. Few can be dealt with by using smelter wastes buried in fertilizers and blackboard chalk encapsuled as human food supplements. All bow to the presence of ocean solids.

The full range of hunger signs cannot be dealt within the purview of this book on ocean solids — it would take hundreds of pages to cover as many crops. One example should do: boron.

Boron is required so that calcium can perform its metabolic function. It is essential in growth processes. When present in the cafeteria of refined nutrients, it presents such abnormalities as crooked stem in celery, internal cork in apples, black heart in nuts, potatoes and turnips, yellowing of alfalfa leaves. When boron deficiency is a problem, death of the terminal bud is a common symptom. Lateral buds continue to produce side slots, but terminal buds on these side slots fade away. Rebranching may occur, but the nelli-branch plant will take on the appearance of a rose tree. In cauliflower, heads fail to mature properly and remain small. Reddish-brown areas may become evident. Terminal buds take on a light green color, pale to base, not so pale at the top. Root ends are affected by brown heart, dark spots or by splintering and cracking at the middle — as with potatoes, radishes, and carrots. Boron is required for translocation of sugar. All this and the trace yield is hardly four ounces per 100 bushels of corn.

This dead reckoning is subject to error because the spending habits of boron are dependent on and perhaps governed by a whole raft of other microenzymes. These have not been provided a biography by science, and most of them cannot be applied except through the agency of ocean solids.

This inventory of details on boron is presented more or less as an aside, for the real basis for the trace is enzyme construction. When the physician Maynard Murray was a young man, only a few enzymes were known. Two decades later, the number identified increased exponentially to perhaps 60. By the time of his death in 1982, the number was measured in the hundreds, each demanding a trace mineral key. How enzymes govern life will be covered in Chapter 6.

5

The *In Vivo* Factor

The years that followed the first full-scale ocean solids tests gave Ed Heine and the world a glimpse of an M.D. whose vision proceeded way beyond the parameters of medicine.

"Little did I realize that for these next 29 or 30 years I would share and absorb just a tiny portion of that vision," said Ed Heine two decades after Murray's death. That vital block of salt, the stone hammer, and the coffee mill still cause the farmer to chuckle. He couldn't have known that greater things lay dead ahead. The next year of dabbling with ocean solids made its own suggestions. In field experiments, there are always special reservations that elude tabulations and rarely earn a footnote. Shortly after fertilizer made from ocean solids had been incorporated into the soil, a 4-inch rain soaked the fields. It came on so rapidly that heavy runoff could not be avoided. Minor rises and valleys escorted the water along with some eroded soil onto a neighbor's land. Some of the water reached a ditch beside a gravel road. The neighbor informed Heine that obviously there would be salt damage because of the fool experiment with a known toxic material. He expected Dr. Murray to compensate him as soon as the extent of the damage was known.

The area into which erosion spilled was a second-year heifer pasture, part grass and part alfalfa. The neighbor watched the sit-

uation almost daily to determine whether the pasture was affected. There was no damage that he could see.

Several weeks later, Heine remembered, the neighbor put his
growing heifers into the pasture. After the animals had grazed
lightly on the field, they concentrated their feeding on the low
spot where soil erosion had given the field an unexpected treatment. Normally, the heifers concentrated their grazing on the
high ground, using the low ground later on. Now came an eye
opener – animals preferred the part of the pasture treated by
ocean solids.

Throughout the summer, this neighbor drove his milking
cows down the gravel road to a daytime cow pasture on the
opposite side of his farm. This practice was routine and time-
honored. Walking the cows along the road had never been a
problem; not before the arrival of ocean solids in the neighborhood; certainly not before that heavy rain soaked the ditch with
some 90 nutritional elements.

"That cow herd was like a 4-H project," said Ed Heine. "He
now had 'club cows' instead of his regular dairy herd." He had to
club his cows, Heine recalled with a smile, past the large area of
gravel road where the 4-inch rainstorm had flooded the ditch
with melted ocean solids. Every time the herd came to this area,
they would stop to eat, defying the herder. Piloting the animals
past the nutrient-rich low spot became a twice-daily challenge,
morning and evening.

Maynard Murray often remarked that so-called lower forms
of life exhibited more nutritional sense than Mensa-rated *Homo
sapiens.* There were times when he expected cattle to smell
before they tested. This did not seem to be the case.

Perhaps some wavelength in the infrared part of the spectrum flashed a signal the bovine could sense; instincts long dor-
mant for the head of the biotic pyramid, namely humans.

Under the shade trees during those weekly visits, Murray
shared insights and talked about experiments he wanted to do.
Some were simple in the extreme. His goal was always a frontal
assault on cancer, osteoporosis, arthritis, and all other types of
degenerative metabolic diseases. Reluctantly, he considers the
psychology of fellow human beings.

One of the 11 investors in Murray's project did what amateurs often do. He went off half-cocked. Watching progress on the Heine farm, this railroader-turned-farmer figured he would order up his own carload of nutrients. He did not know that Maynard Murray had a norm for selecting ocean solids. They had to contain some eight minerals, the type that erode into the air after evaporation had accomplished its chore. Murray didn't tell anyone what these minerals were, but certainly they had to include the four that scientists can't put in artificial seawater — cobalt, cadmium, chromium and tin. The full secret perished with Murray's papers and an unpublished manuscript that was in press at the time of his death.

The errant investor apparently believed any ocean solids would do. He spread his carload on his own land and ruined all the treated crops. As a certified globe-trotter, Murray knew where the deposits were, how to measure the balance, and how to configure deposits as a granular form. Drying them out is necessary to keep the next high tide from returning the undried slush. In this case, financial assistance for future projects was lost.

Ed Heine's assessment of the incident came back to Dr. Murray himself. Henceforth, Murray kept his own counsel. He would reveal all when he was ready, and not to accommodate the greed of betrayal. During those shade-tree visits with his associates, Maynard Murray talked about some minerals not being available to plants, whereas others were present in abundance. Soils toxified by selenium, the kind that laid low Custer's horses at Fort Randall in what was then the Dakota territory, were of particular interest. He explained that a cow consuming only forage grown on such soils would die in six months. A man who ate all of his bread and cereal, cereal made from wheat grown on such soil, would lose his hair in about two years. Maynard Murray farmed out a selenium study to a university — "I forget which one," Ed Heine said — using ocean solids containing selenium. Applied to the land, the research farm's winter wheat harvest had only a modest protein level. It was a tolerable level. Once the door was kicked open, he hoped to enlarge his inquiry, but once reserves dried up, so did inventory studies. The grant money the university demanded was out of the question.

No doubt Murray would have chased down the metes and bounds of all 90 ocean minerals had funds been forthcoming. He hypothesized that if a population in a defined area consumed only food from nutritionalized acres, hospitals in that area would seldom be needed. Without ever meeting the man, Maynard Murray followed in the footsteps of Albert Carter Savage, the author of *Mineralization: Will It Reach You in Time?* Savage studied minerals embedded in Kentucky shale. These were minerals placed on deposit eons ago, ocean minerals diminished by time and erosion when compared to fresh minerals straight from the deep blue sea.

Murray did not believe it was necessarily the lot of human beings to suffer ailments and infections. There was no valid reason for life to come with disease and suffering always tagging along. However, he was quick to add that pharmaceutical and chemical companies would do whatever was necessary to maintain their profit margins.

Some of his medicine had a folklore tint to it, as in the case of a third-degree burn victim. Murray wrapped the burned area in the placenta of an unborn calf taken from a slaughtered cow. Rapid healing resulted, this without a scar in the healed area. Nature presented her own intelligence and recommended against drugs derived from coal tar.

"Countless times he discussed the role of balanced minerals as found in the ocean. There were copper and zinc, silicon, iron, manganese, boron, nickel — the list goes on — some recognized as essential, some still in limbo, yet picked up by plants just the same. Initial tests suggested an uptake of 14, then 18, finally 20 in tomatoes. Grass was the champion. If available in the soil or in a hydroponic solution, grass would pick up approximately 90 elements, and all of them came to be essential if the product was to deliver nutrition and keep its quality." So says Don Jansen, whom you will meet later on in this book.

There are bodies of water that do not have a balance even though the oceans seem to strike a remarkable balance. The Sargasso Sea is one. The oceans off the coasts of Peru and the Baja Peninsula are especially rich, a consequence of thermal vents on the ocean floor that pump out mineral mixes that enrich the waters and nourish native fish stocks. A true balance cannot be achieved in man's time. Whatever the disturbance, total diffusion may take more than any human lifetime. But balance was the order of nature, perhaps the ocean's reason for being.

The power that leaves other minerals in their proper settings resides in sodium chloride. This topic consumed hours of shade-tree time while plants obeyed their biblical injunction to grow, turn inorganic minerals into organic food, and deliver vitamins D and K, among many others. Ocean water and blood are almost as alike as carbon copies. One of Murray's research plots was fertilized with ocean solids, and one fertilized with blood; the visual results were the same when compared to a control area. More research money would have made it possible to do a full-spectrum analysis of the harvested crop following the second year of testing. The results might have been revelatory, since both substances contain an identical amount of ocean water elements, ocean water being inorganic, food being organic. Farm-based experiments with seeds from the previous year's crop were planted "and no hybrid reversion to parent stock was observed," plot manager Ed Heine recalled.

The only occasion the publisher of *Acres U.S.A.* had to tape Maynard Murray came in Overland Park, Kansas, in 1976. He had just completed the manuscript of *Sea Energy Agriculture* with Tom Valentine. The extracts that follow have been pulled from that tape. Ed Heine and Don Jansen confirm that this material encompasses many of the things Murray spoke of during their own Socratic sessions with him under shade trees or at the edges of hydroponic beds.

> Life is contained in a cell. It is contained in a definite value, unlike inorganic substances. In turn there are cells within the cell. (Cell biologists tell us that once a virus inhabits a warm-blooded body, it cannot be extinguished, although it can be controlled. This is hotly debated by aficionados of H_2O_2 and practitioners who introduce ozone into the bloodstream.)

Living tissue has to get its form by either concentrating or diluting inputs from its environment or incorporating the environment as part of its tissue in altered form.

All of life is parasitic. One living thing lives on another. The exception to this rule is plant life, which contains chlorophyll or perhaps a purple pigment. There are actually three pigments by which plant cells can synthesize their own tissue out of simple inorganic things. These are chlorophyll, the pigment in blue-green algae, and the pigment in the retina of the eye.

The latter, with the aid of light, can synthesize vitamins, proteins, etc., out of inorganic materials.

Green plants will not use organic molecules. That is why I question the term "organic." It is a case of doing the right thing with the wrong nomenclature. The farmer and gardener is not feeding the plant organic material. Such nutrients have to be broken down into inorganic nutrients before the green plant can use them.

Nevertheless, all of the relevant organic materials should be kept in the soil, where they can be broken down by bacteria and fungi into an inorganic form for the plant.

Plants cannot use organic elements, and animals can't or shouldn't use inorganic elements or compounds.

Ordinary table salt is the only inorganic compound we consume with impunity. Yet salt is a toxic material, sodium chloride. It causes tissue swelling. This is the reason doctors take patients off salt if they have heart disease or are pregnant. The doctor may or may not know why. The reason salt produces swelling is its inorganic nature. As an animal, you can't utilize it. However, if you have sodium and chloride hooked up in carrot juice or in juices of many other vegetables, it can be ingested without any harmful effect.

It is a common medical practice to administer potassium chloride in a salt form. Here the iodine has the opposite effect than it has when it is tied up organically. If you consume organically tied-up iodine, it steps up metabolism. If you have a toxic factor, take potassium salt; it steps down metabolism. Inorganic iodine, however, prevents certain types of goiter. This is probably because the iodine is hooked up organically in small doses by plant life in the intestines.

Animals can tolerate large doses of inorganic salts because the protozoa and bacteria in the stomach tie up the inorganic salts and, in effect, make them organic.

Human beings do the same thing with iron. Ferric chloride is an inorganic iron. It does not deliver any benefit before it is absorbed by bacteria in the intestine, then released on as an organic tie-up. Therefore, it can be utilized. Failure to achieve such a tie-up can result in either hemochromatosis or Wilson's disease. Animals must have organic elements, either fed outright or complexed by the microorganisms of the stomach or gut. Failure of formative juices or bacterial life in the intestines renders the mineral fix either worthless or marginal.

Life on Earth started in the sea. In fact, human blood is about 25 percent seawater. The trace elements in blood plasma have about the same chemical composition as one-fourth strength seawater.

Fully 85 percent of the life on Earth is in the sea. Life will end in the sea unless man intervenes with atomic destruction. There is a reason for this. The sea receives all the elements washed off the land. Nutrition leaves the land in significant but not incalculable amounts; it has been measured.

The sea is either neutral or a bit on the alkaline side. Consequently, there are two elements that will not stay in solution: phosphorus and iron.

Phosphorus will be the longevity element for life on Earth. For this reason, life is dying at a tremendous rate as phosphorus leaves the land and heads for the ocean.

Admittedly, there are great phosphate deposits in Florida, but phosphorus has to be dispersed and put into solution. Phosphorus forms salts easily with iron and some other elements. In an alkaline solution, it leaves solution and becomes an invisible salt. This is what takes place in the ocean. Therefore, phosphorus is lost to the sea in tremendous amounts and is not recovered except via bird droppings, and then only 1-3 percent is recovered. The loss of phosphorus to irrigation, to erosion and to water amounts to 3,500,000 tons a year.

One percent of living tissue consists of phosphorus. This means that 3 million tons of tissue dies each year and is lost to land use. It goes into the sea and is irretrievably lost unless a way is found to recover insoluble phosphorus compounds off the ocean floor. As one of 90 elements, phosphorus is pivotal.

I turned to the use of ocean solids as a fertilizer for reasons stated, and for some still to be examined. I do not take out the sodium chloride. I use all the traces found in the universe. I spread them on the soil. I have used these solids from 250 pounds per acre to as high as 2,200 pounds per acre. I also grow my crops in hydroponic solution using ocean water.

To the ocean water we have to add some N, P and K because land plants have become acclimated to a higher concentration of nitrogen, phosphorus and potassium than is found in ocean water.

Whereas phosphorus does not stay in seawater, nitrogen is fixed in that medium. Fixing nitrogen out of the air by electrolysis is nasty compared to what nature does with bacteria. Lightning also fixes nitrogen. Still, bacteria in the sea are the main source of nitrogen fixation. When you use sea solids on row crop acres or pastures, bacteria — Azotobacter and Clostridium, that is — fix nitrogen out of the air, using ocean water as a food supply. By using these solids on the soil, you gradually build up nitrogen in the soil. Finally, you'll get to a place where you do not need outside nitrogen. That is what is meant by having the natural nitrogen cycle working. There the soil bacteria perform exactly as they do in the sea.

About three-quarters of the Earth is sea. For reasons already stated, I failed to find sea creatures with diabetes, arthritis, cancer, or malnutrition. There are no hospitals on 72 percent of the Earth.

My objective was to grow plants that were better than the norm. After working out the dilution by trial and error, I found I could.

In the late 1960s, corn blight destroyed most of the nation's corn production. Yet the corn I and my associates grew that year was absolutely immune to corn blight; this to the very row in which sea solids had been applied. The same proved true for corn smut, a corn fungus. We killed off virus diseases in other crops just as easily, mostly tobacco mosaic virus and tomato mosaic virus.

This immunity was also confirmed in all cases with trees. Fruit trees with curly leaf repaired themselves when sea solids were watered in. Fungal diseases bowed to the efficiency of sea solids. Corn rust faded.

Center rot in turnips is a bacterial infection. This one is caused by staph, the author of boils, staph pneumonia. Immunity to staph infection can be built up — viral and fungal — by using sea solids.

What happens when we feed ocean solids to animals? Feeding oats, corn, and other grain to animals, changes become evident. I have experimented using animals with cancer. Grains grown with ocean solids literally feed cancer out of the C3H mice (bred to be cancer prone) in two generations. Cancers were cut from 97 percent to 55 percent in the first generation. Each generation revealed a drop in the cancer rate.

Avian luekosis in chickens has been brought under control the same way. Sarcoma in chicken kills rapidly, usually in five days. In this experiment, feeding grains with ocean solids failed.

Arthritis in rats evaporated via experiments with ocean solids. Nutrition is effective as a remedy, but even more effective as a preventive.

The farm is the beginning of prevention. Cows, pigs, all animals respond somewhat the same.

There were animals that puzzled more than they enlightened. For instance, gray horses almost always develop and die of cancer. Black horses never do. The bigger the horse, the sooner it dies of cancer, of certain types of melanoma. This is a pigment- producing cancer usually associated with birthmarks in human beings or a black mole of the kind a gray horse gets.

Blood analyses record minute differences in the amount of manganese in gray and black horses. This may explain why one gets cancer, the other not. (Manganese has been implicated in inhibiting the uptake of copper, a precursor to symptoms of encephalopathy, Mad Cow disease, Creutzfeldt-Jakob disease in humans.)

Gray hair has less silver and less manganese than normally colored hair.

The wisdom of Maynard Murray was not exhausted before the tape was gone. He was planning a 1,000-child study feeding 50 percent with sea solid-grown food, 50 percent regular base. Colds, IQ and all the considerations that extension field study could account for were to be codified. The study was prompted by results in animals where it was observed intelligence improved as a consequence of feeding food that was ocean grown. In one experiment, Murray injected DNA into the muscles of an animal, then extracted that material and delivered it by injection into another animal.

Finally, it was aging that commanded attention during Dr. Murray's final years. He reasoned that it was due to a dilution of trace minerals in the system. Wrinkles in the skin means dilution of zinc. Sulfur dilution exhibits itself in senility.

Cell reproduction posed at least as many questions as it answered. "If I take a piece of my tissue and put it in a culture, I can grow it. It will divide 50 times. One cell produces two, two to four, etc. At the end of 50 times, regardless of what I can do, reproduction ceases. This varies with animals. A mouse doubles 17 times, a rat 23, a human being 50," Dr. Murray said.

Strangely, this happens on land, not in the ocean. A sperm whale 60 to 100 years of age will have cells that keep on multiplying. All warm-blooded animals in the sea will do the same. All cold-blooded animal life in the sea will elude cancer unless it comes too close to shore. Isolated in a tissue culture, a cold-blooded sea creature's cells keep right on doubling.

Maynard Murray's own journals describe the range of interests he followed until they took control of his life and profession:

In one experiment, 24 rabbits were obtained. Twelve were designated experimental and fed with plants grown on sea solids, while the remaining 12 were labeled controls and fed accordingly. All of the rabbits were given a high-cholesterol diet for six months, which produces hardening of the arteries. The control group did develop hardening of the arteries, and all had died within 10 months. The experimental group did not exhibit hardening of the arteries.

A breed of rats that developed a disease of the eye was obtained. The 10 that were put on experimental food showed no deterioration of the eyes and bred five litters. Those on the control food diet all died secondarily of eye disease.

Hay was grown in Lennox, Massachusetts, on soil fertilized with 2,200 pounds of complete sea solids. Corn and oats grown in Ohio and Illinois on soil treated with complete sea solids were also obtained and fed by a dairyman to pregnant cows. One of the problems previously experienced by the dairyman was that his newborn calves from these purebred cattle had difficulty standing in order to nurse when they were first born. They often had to be held for their first nursings and were often not uniform in size. However, when calves were born from the cows that had been on food grown on complete sea solids-fertilized soil, all of the calves were immediately able to stand up to nurse and were uniform in size.

In 1970 an experiment was conducted in southern Wisconsin, the report of which follows:

A 40-acre field on which corn had been grown for the preceding nine years was treated in 1969. Although a portion of the field required three tons of lime per acre, we applied four tons. In the spring of 1970, 110 pounds of anhydrous ammonia was added to the entire field, followed by an application of sea solids to 14 of the acres, which constituted the experimental segment. The rows were 80 rods long, with each plot about an acre, or 10 rows, in size. Two hundred pounds of sea solids were placed on the first acre. Each of the remaining acres had an additional 100 pounds poured on it than the one before, so that the last acre had been enriched with 1,500 pounds of solids.

King Cross Cord PX-610 was planted along with 150 pounds per acre of 6-24-24 starting fertilizer. Germination of all corn was excellent. Throughout the growing season, the crop showed a stair-step effect in growth, with corn on the 1,500 pounds of sea solids per acre showing the best advancement. All plots except for the one exhibited corn blight, marked by fallen stocks, and a difference could easily be noted in the effect of the blight, starting with the 400-pound plot and diminishing with the 1,500-pound acre.

The cord was harvested in each test plot on November 7, 1970, by one round with a picker to determine how well it shucked, and to determine blight damage to cobs and kernels. The remainder was combined. The 1,500-pound corn yielded 154 bushels per acre, while the untreated acre yielded 115. The yield increased as the sea solids increased. The weight of the 1,500-pound corn was 57.5 pounds per bushel compared to 53.5 pounds per bushel for the untreated acre. The 1,500-pound corn consisted of 20 percent moisture, while the untreated consisted of 25 percent. The treated corn leaves were much greener at harvest, even though the corn was less moist.

There was some evidence of corn blight on the 1,500-pound corn leaves, but it did not affect the ears. The cobs shelled out with complete, whole kernels, and the cob was solid. The untreated and low-application (100 and 200 pound) corn suffered ears with rot at the ends of the cob.

This same seed corn (described above) and fertilizer treatment was planted in another field that had been used the previous year to grow alfalfa. Manure was spread, and the alfalfa field was plowed under to prepare for the seed corn planting. Although no sea solids were administered, the yield on this plot was 130 bushels per acre. However, the best corn yield, weight and moisture on the 3,000-acre farm was the corn grown on the acre that received 1,500 pounds of sea solids.

Steers weighing 1,100 pounds that had been fed regular corn were fed on corn grown on the 1,500-pounds-per-acre plot described above. The steers were fed until they weighed up to 1,400 pounds, using one-third less corn than previously required with regular corn, and they appeared to be in very good condition.

Maynard Murray summarized:

"There is a very real and pressing need to develop an experimental animal that has consistent chemistry. I suggest we raise "mini-pigs" on food grown on sea solids-fertilized soil. Pig physiology is much closer to human physiology than normal laboratory animals such as dogs, rats, mice, guinea pigs and rabbits. If we develop this breed of animal with muscle tissue, nerves, lungs, heart, kidneys, etc., that are characterized by constant body chemistry, we then will have an animal that ideally lends itself to very precise research. This would aid existing efforts in testing drugs, antibiotics and disease resistance in general.

"Of equal importance and equal urgency is the need to conduct experiments to see what the long-term effects food grown on sea solids will have on human beings. We must recycle the sea – and now – for our health and the health of future generations."

Yet freshwater trout in Minnesota do have cancer of the liver. Their tissue stops doubling in a culture medium at 23 to 27 times.

The pattern described above applies to all infectious diseases incubated by viruses, bacteria and fungi.

Some 92 million elements of ocean solids are deposited in shallow ponds twice a year off the Sea of Cortez. This mother lode of nutrients is replenished each time the semiannual master tide rolls in. Only the Moon's orbit and Earth's trip around the sun regulate these phenomena.

Used in wheat fields in selenium or molybdenum excess areas, sea solids seem to govern uptake of the excess, making the wheat commercially useable on otherwise condemned acres. When a plant has everything it needs, it will not take up toxic amounts or excesses found on land or even solution.

In growing tomatoes, Dr. Murray measured 94 to 104 percent vitamin C, six times the amount available in commercial crops. Trace elements, 18 percent more uptake! Occasionally in conversation Maynard Murray returned to the high seas, where depth is measured in fathoms and speed in knots. It was aboard ocean-going trawlers that he examined life forms caught by nets. He was a young man then, a recent medical school graduate, and the Great Depression hung over the land. In the West, dusts of the soil boiled up into great clouds, some of which blew out over the Atlantic Ocean. During his medical school years, Murray often referred to experiments published in the medical literature that subjected land creatures to cancer. He found he could not induce cancer into toads. Why not this life form? Other species readily absorbed injected cancer. It was this simple fact, and a clever remark from a researcher about cancer being absent in the ocean, that made Murray take to the sea.

He was determined to prove it true or false, he later reminisced. He was not motivated by folklore or third-hand reports. Those eight months of on-ship research stitched the dots together and gave Maynard Murray a picture few physicians could see.

A transfer of disease conditions inherent to plant life seemed inevitable. He encountered something that bordered on shocking when life − plant, mammal or fish − came aboard. Aging as we know it simply did not occur in these animals. Small whales became great whales, but their tissues and organs were as pristine as those of a baby. Sailors who had spent a lifetime finding fish told the physician this or that turtle was 200 years old, but the untrained eye couldn't tell. One turtle brought aboard was an estimated 400 years old. As he examined the veins and blood vessels of these ancient creatures, the flexibility he found once again reminded him of a newborn. Fish programmed to go upstream to spawn developed hardening of the arteries and sunken cheeks, all within a few weeks after going into fresh water. Ocean salmon swimming up the Columbia River were a good example, Murray told Ed Heine. It excited the doctor to learn that a change so dramatic could occur in such a short time. It also ushered in an awareness of the ocean's natural balance, a condition that had no parallel on land. There was nothing consistent about land, country to country, state to state. Nor did the addition of only several minerals to the fertilizer mix rectify the shortfalls. Minerals not present in equilibrium complex and confuse each other. Often they telescope themselves into excesses and thus install disease anomalies into plants and into animals that eat these plants. The tawdry practice of using smelter wastes as feed supplements answered a disposal problem, but it also delivered the most devastating of all diseases, cancer.

The land's loss added to the sea via a watering process at least as long as it took Mount Everest to rise out of the ocean to become the highest mountain on planet Earth. The equation presented itself. If the land's mineral mix came from the ocean in the first place and was not returning, then the art of fertilization depends on the ocean for its continuing nutrient fix. Extrapolation becomes a condition. Ill health is nothing but an imbalance finding expression in signs and symptoms. The organisms that Pasteur assigned such a lethal role are no more than nature's cleanup crew, there to attack when the terrain invites them, helpless when immunity sets up a barrier. Cure cancer? Why so much attention to curing cancer and so little attention to preventing this disease anomaly?

Maynard Murray never lost track of his ocean odyssey's fortuitous timing, well before World War II. The objects of his observations were timeless. But by the end of the war, the planet was changed forever.

Over 40 years intervened between Murray's early notes on the sea and the great physician's death. During that time, B-29s dropped atomic devices on the crowded cities of Hiroshima and Nagasaki, setting in motion a series of events that would spread a new form of pollution across the globe. Cancer, always present, suddenly became a nutrient sponge. At the same time as man-made radiation entered the environment, a new agriculture was touted.

"We are rightly appalled by the genetic effect of radiation," wrote Rachel Carson in *Silent Spring*. "How then can we remain so indifferent to the same effect produced by farm chemicals used widely in our environment!" Maynard Murray was cognizant of every development in science, especially in human medicine and health maintenance. On April 25, 1953, *Nature* magazine published a model of DNA, based on the work of Maurice Wilkens and Roselyn Franklin's X-ray photography.

DNA is a blueprint of sorts. It tells cells how to divide and reproduce copies of themselves. Picture a twisted rope ladder. All DNA structures are shaped this way – on a flower, a dog, a human being.

The rungs of the ladder are made up of four components – adenine, cytosine, guanine and thymine. These are usually written as A, C, G and T. A can only pair with T, and C with G. Base pairs reproduce themselves, and that is where genetic manipulation enters the scene. Millions of these base pairs form genes. Evolution has taken up the chore of detecting the base pair reproduction, frequently and even usually improving the life structure. Genetic engineers have learned how to add and delete from the ladder. They came by this knowledge in the following way.

An Austrian monk named Gregor Mendel hinted at the direction for research, but it remained for Frederich Griffith, a British microbiologist, to discern what was later identified as DNA. In the early 1950s, Maurice Wilkins and Rosalind Franklin at the King's College in England used X-ray refraction photography to study DNA. They came up with an outline of the DNA molecule. The Nobel Prize would have gone to Franklin, except for her untimely death. Nobel Prizes are awarded only to living persons. Using Franklin's work as a background, James Watson and Francis Crick, working at Cambridge, constructed a model of DNA, the double helix. In 1953, Wilkins joined Watson and Crick in receiving the Nobel Prize.

It was a small step to discover naturally occurring enzymes that act like molecular scissors for the purpose of adding or deleting rungs from the DNA ladder.

Breaking the molecule has been applauded because of the potential for fighting hereditary disease conditions. Use of the technology has permitted laboratories to manufacture things like insulin. Thus was born the idea of cutting and recombining at the molecular level. Thus was born the idea of finding a trait in one organism and transferring it to another organism. Thus also was born the idea of engineering the totality of life.

When Watson and Crick proposed a double helix model for the study of DNA, their report was described as follows: "This structure has novel factors which are of considerable biological interest." It was perhaps the understatement of the century. It came at the time Maynard Murray decided to find out what it was that impeded cell reproduction, especially cells in wild proliferation, namely cancer. This opening of an understanding of molecular biology and genetics merely ratified the nutritional thesis Murray had championed all along. As a preliminary understanding of the process used to create amino acids and proteins, this DNA ladder would ultimately lead to gene therapy and hammer home the truth that we are what we eat. DNA has now become a metaphor for information-bearing genes, genes that are offended by radiation, farm chemicals and malnutrition. Murray wanted an explanation for life and death. The molecules of DNA are held together by phosphorus. Phosphorus compounds supply the energy for muscle in the human body. The discovery of a molecular basis for life delivered a law, not a hypothesis.

As the sun settled toward the west one day, a grandiose idea came to Dr. Murray – the sea must be recycled. Why not? Hadn't the continents been recycled? The idea of continental drift and tectonic plates had been around for at least half a century. But it wasn't until the 1960s that the insight of a visionary became settled science. Dr. Murray had a long way to go before recycling the ocean became settled science.

6

Seaponics Farming

Thirty years of research and observation cannot be summarized in one whiplash line. Those plants in Dr. Maynard Murray's basement had messages to deliver, if only humans could read them. At the time Murray set out to prove or disprove his theories, a man named Clive Backster attached a lie detector to a hibiscus plant. There was an electrical reading because, after all, plant life and human life are electrical. When a worker entered the laboratory to offend one branch of the plant by smashing it between his fingers, the plant must have taken note. The worker left the room, then re-entered, the lie detector went crazy. Murray did not have the reserves to follow every hunch, but he could speculate on the higher screams plants would express if they could talk.

Many of Murray's experiments were with produce he was growing hydroponically, feeding them ocean solids rather than N, P and K commercial fertilizers. The earliest results were strong plants and an absence of plant diseases. Not least, the produce tasted better. There were numerous Chicago-based blind taste tests that bore this out. Since fungal and insect spray destroyers did not torment the hydroponic trays, there was no need for rescue chemistry. Weeds didn't figure.

The basement hydroponic base soon became a small commercial one. He even sold produce to a few Chicago food stores. In 1958, he purchased several acres of farmland in Florida to increase production. His three-acre tract soon had 178 hyroponic beds, each 100 feet long, 4 feet wide. Tomatoes and cucumbers were grown by the ton. Offered to his farmer counterparts, his findings were met with a gale of silence.

He came up with a commercial name, Seaponics Farming.

That animals knew the difference was established quite early, both when feeding ocean solids-raised grain and when lacing bunk feeds with a measure of those same ocean solids. "My animals just plain like the sea solids," summarized Ed Heine years after the last experiments were history and Dr. Murray's conduit for getting the stuff from the ocean direct. "They would choose sea solids-raised grain over conventional grain when given the choice. They can sense the difference. The bottom line recorded that it took less feed to produce a pound of beef, higher quality beef with superb taste, and better consistency."

Much the same was true for produce from the greenhouse. The same general findings became the legacy of poultry trials in which chickens were fed ocean-grown grains and the crystals themselves in the ration. They not only grew faster, they delivered larger eggs. Dressed out, their meat was a better consistency.

Studies and statistics stacked up like bricks on a work site, and still regulation faltered.

Dr. Paul Dudley White, President Eisenhower's personal heart physician, heard about Maynard Murray's work with sea animals. Unfortunately, White passed away before he could pursue the matter.

There is such a thing as getting a break. Roselyn Franklin died before Nobel laureates were passed out in connection with the DNA, and she remains unrecognized.

Maynard Murray carried on alone with his "investors" gone and only a polio victim as a research helper.

In 1963, Dr. Maynard Murray expanded his research to include mice, C3H laboratory mice. This is a strain developed to incubate breast cancer.

His trial called for one group to be fed grain grown on ocean solids. A control group received regular cereal grains.

"Normally, C3H mice have a cancer rate of at least 90 percent," Dr. Murray summarized. "That's the level one control group hit. The experimental group rate was only 55 percent. And the rate for the second generation of this experimental group of parents fed sea solids-raised food dropped to 2 percent. Murray proved that he could literally feed cancer out of the mice in one generation.

Maynard Murray didn't stop with this amazing result. He applied lessons learned to human population areas with low cancer rates. Such areas always had soils rich in multiple trace nutrients. This study was made before the debauchery of agriculture was complete, before most vegetables were grown in four states for the entire country, before debased meat protein production in feed lots became the indescribable norm.

In the main, however, it was the greenhouse that had the attention of this physician. Using ocean solids, he could bypass soil, which in any case was deficient. River gravel could anchor the plants, and routine flushes of ocean solids water seemed to make an end run around a century of settled knowledge on exchange capacity, phosphate beds, pH, even that absolute requirement for major nutrients. The old Albrecht formula reserved only 5 percent of the soil colloid for trace nutrients. Maynard Murray's research in effect pronounced the traces the most important.

If the medical profession didn't listen, then neither did agriculture. It was an age of Poison Control Centers, Extension advice to get bigger, more efficient, or get out. The underground that pursued organics, biodynamics, or simply went their own way – as did Harold Simpson, the Chicago nutritionist who suggested therapeutic food – all were considered curious folks who simply weren't in the mainstream. It would be well into Century 21 before even a token number of farmers took seriously an absence of trace minerals in the soil, even longer for the population at large to make the connection between trace nutrients and enzyme production. Henry Schroeder's *The Trace Elements and Man* languished as unsaleable merchandise in the bookstores.

By the year 2000, Dr. Edward Howell's *Enzyme Nutrition* penetrated the stonewall with logic and clout. He was standing on the shoulders of giants, Schroeder and Maynard Murray included.

"If the pace of discovery is maintained, we'll discover 25 new enzymes every 50 years."

Without losing track of the fact that all enzymes have to have trace mineral keys, any rundowns on the mysteries of enzymes must start with still another element, oxygen.

Oxygen is essential for life. The term "essential" usually puts "regulated" on the other side of the equals sign. In fact, cells and components of cells can sustain damage if there is an excess of oxygen. O_2 is the symbol for the oxygen molecule.

The "2" indicates that there are two oxygen atoms bound together. It is an axiom in chemistry that when atoms interact with each other, there is a chemical reaction that generates free radicals. According to settled science, atoms have electrical particles called electrons. These circle the nucleus. In turn, the atom is in a favorable state when its electrons are paired.

The O_2 molecule has an equal number of shared molecules. The moment one electron is pulled away from the molecule, the atom with the unpaired electron is out of balance. It seeks another electron to pair with and will rob it whenever it can. A negative-seeking atom or molecule is called a free radical. Singlet oxygen (O_2 minus one electron) is a highly reactive form of oxygen. Singlet oxygen is called oxidated oxygen.

There are other free radicals, such as hydrogen peroxide, which has two unpaired electrons, and hydroxyl, with three unpaired electrons. The species are called superoxides.

All are capable of damaging the DNA, proteins and other molecules in the cells, a consequence of oxidation, the so-called oxidation reaction.

Free radicals so produced are called oxidants. Antioxidants cancel out oxidants. They do this by handing over the electrons the oxidants need, thereby preventing damage. Vitamin C, the vitamin E family, cartenoids, beta-carotene — all have this role.

Free radicals are not all bad. They are produced by immune system cells. They engulf bacteria and pathogens when they invade our bodies.

These cells use superoxides and other free radicals for the actual annihilation of bacteria. They keep these reactive species sequestered in tiny cell sacs. Under these circumstances, the free radicals will not harm other parts of the cell or buffer inactivation by antioxidants.

Free radicals, much like everything else in the human or animal body, are like a two-edged blade, either protective or damaging the host. The governing factor is balance.

It goes without saying that internal toxicity creates an excess of free radicals. So do smoke, irradiation and farm chemicals, these from the outside. The damage accounted for by cancer therapy should be at once apparent. Even the junk-food diet can hand over a dose of free radicals. The damage from microwave food can be imagined. The DNA becomes a target. Mutations can result.

The pregnant woman assaulted by too many free radicals can produce a mentally retarded or deformed child.

The diseases that result fill a medical dictionary, everything from cancer to cataracts. Even aging is a consequence of free radicals operating in aerosol fashion. DNA damage has been indicted in the failure of a cell to copy itself. Cells that must divide fall prey to the invader.

Much like soil, the human body forgives insults – up to a point. Continued insult overwhelms the system, causing cancer. The common meaning of the term departs from the metabolic meaning, for, in a manner of speaking, all degenerative metabolic diseases are cancer.

Oxygen comes off as the great healer. But this does not mean that too much is better.

The human body cannot forgive anything that impedes enzymes. The opening words of Dr. Edward Howell's monumental study tells us as much.

"The length of life is universally proportional to the rate of exhaustion of the enzyme potential of an organism. The increased use of food enzymes promotes a decreased rate of exhaustion of the enzyme potential."

This is Dr. Howell's enzyme nutrition axiom, and it explains why basketball players drop dead on the floor, why marathon runners collapse and die at times, and why people who do not buffer cooked food with fresh fruits and vegetables can count on debilitation and a shortened lifetime.

It was the consideration of enzymes that took the ocean solids story in a new direction, one that will unfold in the several chapters that follow. The trace mineral requirement known as the enzyme key has been stated. The assortment in terms of the state of the arts is now so vast an auditor would be required to codify an up-to-date inventory, and that inventory would be obsolete the moment it was tabulated. There are some few general principles that must detain us just the same.

Howell holds to the proposition that living organisms are spark-plugged by an energy form quite distinct from the calories we ingest via the agency of enzyme action. Enzymes are not merely esoteric medical experiments for brain function, speech, action, aggression or retreat. They are an essential life force without which we cannot drum a finger or enjoy an esoteric moment. The enzyme is, biology speaking, not dead chemistry or caloric heat.

This insight destroys the conventional ignorance that has presented itself as settled science for at least a century. It argues that the enzyme is not merely a catalyst that functions like a wrench or a tire, doing its work and never entering the biological transaction, therefore never working out. Indeed, the enzymes are used up by the process of living. The tip of energy on which enzymes rely is seated in the substrate, nearing the substance being neutralized.

Even the most hidebound chemist concedes that it takes microorganisms to build enzymes, but they seem to hold enzymes as merely backup, to borrow a common phrase, merely expendables, like soldiers in the great battle.

This is not the state-of-the-art opinion, but it is one we respect as we make the case for broad-spectrum applications of sea solid fertilization to all forms of farm and hydroponic production. Like solids are the living forms they sustain, enzymes are quite exhaustible.

Enzymes contain a protein and vitamin payload. These in turn serve the enzymes of which they are a consistent part.

The enzyme spectrum builds itself as our examination continues. It notes quite correctly that heat disrupts enzymes, nutritional heat that is. Radar ranges annihilate enzymes the way a sledgehammer crushes a ping pong ball, and commercial irradiation and preservatives are molecular wrecking balls swinging in arcs that would have to be measured in nanoseconds of time per swing. The list of infirmities caused by an enzyme shortfall fill encyclopedias and *Physician's Desk Reference* with small print.

Howell holds that the dead-food diet available to *Homo sapiens* enables the human machine to specialize in breeding diseases, creating a bodily terrain that invites infection and gives it room and board. Measured against the life terms allowed wild animals, human life is not healthy, brutish and short. Diseases of animals can be summarized in a few pages. Most of them are a consequence of malnutrition and bad husbandry. The near total sweep Johne's disease has imposed on the Continent is one good horrible example.

In *Eco-Farm, An Acres U.S.A. Primer,* it was settled without fear of contradiction that plants in touch with balanced nutrition, with a full variety of trace minerals, create their own hormone and enzyme potentials and therefore protect themselves against bacterial, fungal, viral and insect attack. Much the same is true for human beings.

State-of-the-art research tells us there are three types of trace mineral-keyed enzymes: metabolic, digestive and food enzymes. The enzyme starts digestion because it is contained in the food itself. It handles nature's work unless annihilated in the food separation process. All the packaged food on the grocery shelf is enzyme dead. Where food enzymes fail, which is the case most of the time, digestive enzymes step in. The burden is awesome. The cow with its four stomachs, a 40-gallon digestive vat, or the whale's extra chamber are food-holding pantries that can deal with some nutritional shortfalls because they make maximum use of food enzymes, but these blessings are not made available to mankind.

It can be stated that every symptom in the human body is governed by metabolic enzymes. The task is awesome. Proteins and fats and carbohydrates have to be built into feeding functions. This is why the enzyme art goes up with almost every research day, as even more of nature's secrets are recorded.

We are told that there are 98 enzyme workers on the job in the arteries alone. Much the same is true for every organ, every muscle, every digestive choice, even every thought coming out. Each has required trace nutrients needed for this wonderment. Accounting for a unit-by-unit balance is something akin to counting the grains of sand on an ocean beach. It beguiles the imagination to think of the enzymes required to operate the heart, brain, kidneys, lungs, spleen, even the small and big toe.

Digestion is key to metabolic enzyme production, as trace nutrients are key to all enzymes. Proteases are the enzymes that enable protein digestion. Amylases handle carbohydrates. Lipases digest fats. Foods themselves fight to enable digestion by giving digestive enzymes an assist. It is this consideration that has taken the ocean solids story into a new venue, one that will unfold with clarity in the final chapters of this book.

For now it is enough to point out that the availability of food enzymes enables digestive enzymes in the law of adaptive secretion of the digestion factor. It is the absence of food enzymes that overbooks digestive enzymes and overbooks the draw on metabolic enzymes. The only remedy for the nutritional bankruptcy facing almost 100 percent of Americans is food capable of uptaking minerals, and minerals being in place for grown food to uptake. In Howell's opinion, the metabolic checking account becomes overdrawn for the above-stated reasons. The problem is exacerbated when feedlots subject the body to marathon drains, as in the Boston Marathon, the nonstop soccer game, the linebacker clash with the right tackle in football. Deficiency without replenishment is a one-way ticket to metabolic disaster.

When an excessive metabolic drain on enzymes occurs because of digestive enzyme shortfall, then metabolic enzymes falter in doing their body repair work. The cascade of degenerative episodes acts like dominoes falling. It is subtle and lethal, and there is little in the medical armory to repair the symptoms even if detected.

In the early 1930s the supermarket replaced the smaller grocery store. Instantly the name of the game became shelf life, meaning shelf death. Packaged foods were replaced by processed foods, meats, cheeses, baby foods, all designed to wait it out on the shelf until sold without spoiling. All were dead, dead, dead. Often, even insects wouldn't bother the stuff.

The individual supermarket was replaced soon enough by the chain, and the hunt for shelf life continued its insanity. The reasoning led to products that looked like food and could be made to taste like food, then it could be used as food. Nationwide, physical and medical degeneration escalated, cancer exploded, and teratogenic birth became a disgrace to civilized society.

The grant money went to shelf-life research. Even so, the mere 80 known enzymes known in 1930 became 200 by the middle of World War II, 660 by 1957, 850 by 1962, and God only knows the count for the present.

That shelf life and life value would be ordered to clash seemed inevitable. People who are literate in the subject now know that foods should contain their own enzymes to assist in digestion. That is why the raw apple is loaded, the cooked apple empty. When digestive enzymes are required to do the entire job, the strain on metabolic enzymes is unspeakable.

This is why Dr. Maynard Murray took inventory on trace minerals taken up by various vegetables. Tomatoes proved to be the grand champions. They took up 55 or more trace minerals for enzyme construction when grown in ocean solids hydroponically or in soil field trays or garden soil. Other vegetables varied according to species. Wheat grass came close to uptaking almost 92 elements.

Conclusions are always ephemeral. When one layer of the onion's skin is peeled away, there is always another layer. To say that enzymes govern the reduction of diseases is to say we now know some of the questions. This is necessary before the answer can be forthcoming. Even if we come from the sea and rich fertile soil, we have to answer a lot of questions in between. Nature will cure, certainly, if metabolic enzyme activity is in place. Answers beget answers.

Enzyme studies have answered the question that has dodged pasteurization ever since the days of Louis Pasteur. Simply stated, the process kills enzymes in milk – it makes the product next to worthless. Even heated water injures enzymes in foods. It is not hard to comprehend what happens to American youth fed on hamburgers loaded with soybean meal, the soy being high in estrogens and enzyme inhibitors. The drain on the metabolic bank account is one thing, estrogen interference with sexual development is another. Consumption of too much soy makes hostages of pancreatic enzymes, resulting in enlargement of the pancreas and finally impaired health.

The food enzyme equation is best explained by cattle and sheep. Both have four digestive chambers, three of which are devoid of enzymes. These animals depend on the enzymes in raw food to give them an assist. These feeds have to contain an inventory of minerals. Since foods themselves do most of the digestive work, the several animal organs are smaller than their counterparts in human beings, taking role differences in species into consideration. Metabolic enzymes are preserved, and the bank account is depleted continually.

The whale has three stomachs. These are cavernous affairs almost devoid of digestive enzymes. Whales depend on prey to bring along their own enzymes. These fish and porpoises are warehoused in one stomach until needed, their enzyme content no doubt processing with the system until moved along to the next chamber.

Dr. Maynard Murray's scalpel literally cut to the chase while on his ocean quest. As his experiments came to an end, he connected the dots together. They make up the rest of this report.

Peel off another layer of the intellectual onion skin, and now inspect enzymes like a genie out of the bottle. For most of husbandry's history, the crop of a chick was thought to have no function. Likewise, enzyme studies finally revealed the fact that seeds have enzymes as well as inhibitors, one trumping the other according to genes and species.

We share the evolutionary process with all creatures, more or less – even the whale and its prey. There is a peristalsis-free food-enzyme section of the human stomach. It is where food enzymes, if any are present, do their work. Even mastication assists, always boosting enzymes in the food section. Soon trace food enjoys breakdown until acidity closes down digestion, usually in 30 minutes. Now pepsin intervenes. If food intake is not buffered with raw food, 90 minutes pass and nothing happens.

These counteractions and considerations left unstated caused Don Jansen to pick up on the ocean solids story, couple it with the findings of Ann Wigmore, and shake out the results that now amount to these pages.

Now the circle closes. At the beginning of the last century, Casimir Funk unveiled for the world the essential role of vitamins. For some years, numerous investigators debated what they could find out about the role of basic minerals. Then the traces came to the fore, with Maynard Murray calling attention to the only sure-fire source, the ocean itself (really Ponce de Leon's fountain of youth). The traces led to enzymes and the rich bag of knowledge and nutrition they account for. The work of future generations seems indicated. It's to remove "shelf life" from the shelves and to replace it with *life* itself.

Improper enzyme function can install malnutrition in the presence of plenty. Moreover, industrial poisons can annihilate enzyme functions. Included under this heading are preservatives, hormones, heavy metals and inorganic compounds, some of them masquerading as food supplements. Inoculations on the scale used by the military can destroy proper enzyme function, for which reason past war casualties through illness have sometimes run in the thousands, while battlefield casualties were counted in the hundreds. An enzyme shortfall distracts digestion and kicks open the door to exhibit signs and symptoms. When food has an extended residence in the system, healing and regeneration become impossible.

Photo Interlude

An acre of basil grown hydroponically.

An acre of cucumbers and tomatoes grown hydroponically in ocean solution, under a plastic roof.

Corn and watermelon growing in hydroponic beds on Dr. Maynard Murray's farm.

One acre of cucumbers grown hydroponically.

Don Jansen and one acre of okra – grown in 64 hundred-foot beds.

One of two large outfits preparing the soil for wheat planting in the fall on Jansen's farm in Nebraska.

Jansen's Nebraska farm produced 5,000 acres of wheat each year.

Wheat harvest in Nebraska, 1978 — 5,000 acres being worked by 16 combines. Total harvest: 150,000 bushels.

Jansen's buffalo.

Dr. Murray's hydroponic tomatoes, grown in ocean water.

Ocean-grown wheat grass.

The ocean crystal beds in Baja, California, Mexico – no longer usable for ocean-grown purposes.

Ocean Grown's greenhouse.

7

The Ocean Presides

When the *Pequod* went down in the last chapter of *Moby Dick*, in Herman Melville's words, "the great shroud of the sea rolled on as it rolled five thousand years ago." Was this to be the fate of Maynard Murray's life and work? Its summary at the time of his death was clear enough. In addition to the details relayed before, Murray wanted to find a plant that would uptake all the minerals found in the balance of the ocean. At one time he thought it might be achieved with ocean plants actually growing in the sea. He did not consider it possible using land plants. There was, after all, the great soil food web to be considered: "The fungi, the grazers, the cysts, the bacteria, the protozoa." Not only did they rate study and testing, they had their reasons for being, their role in determining whether a plant would live or die, be healthy or disease free, whether it would contain useable nutrients or not.

It takes an incredible inventory of soil life to ready inorganic minerals for plant uptake. The chain begins with one-cell bacteria and graduates up the scale as algae, fungi and protozoa join the web. The soil matrix also includes nematodes, arthropods, earthworms and insects, some friendly, some not. Also, there are small vertebrates and, of course, plant roots. All this work in symbiosis does what the ocean does for the grower who opts for ocean energy agriculture. It is the function of sea life to consume, grow, and migrate through the soil, cleaning the soil and the air in the process. Allowed to live without the interaction of toxicity, these creatures decree health and survival during heat waves and cold snaps. Soils all have their problems. They vary according to the zoological development, rainfall, and the degree to which soil life forms create a natural nitrogen cycle and carbon cycle.

It is the prime function of soil organisms to decompose organic compounds, to separate the carbon from the inorganic elements, to clean up pathogens contained in residue and manures. The unpaid workers of the soil can take anything apart, even the bed of a truck, all in good time. They can dissolve pesticides, albeit not the modern agricultural chemicals installed during the last half-century, at least not in a time frame suitable to mankind. The task of sequestering nitrogen is one of the most important. They fix nitrogen from the atmosphere. Some of the little critters aggregate the soil and fight parasites. Soil depletion goes with age and use. That is why scientist George H. Earp-Thomas discovered it impossible to find cobalt in New Jersey soils. Not only are many of the traces missing in farm soils, they are almost impossible to replace in the absence of ocean mineral availability.

Eat and be eaten is the mandate for sea life. An understanding of the food web is necessary to transport lessons already a matter of record to the world of ocean dusts and ocean-grown plants. The flow of nutrients tells us that nature is too complex for computers and even man's puny brain. As we understand it, however, the primary producer fuels the web. Plants, lichens, mosses, photosynthetic bacteria and algaes all use the sun's energy to fix CO_2 from the atmosphere. Other soil organisms get their energy from consumption of organic compounds found in plants and water. There are autotrophs that live on nitrogen, sulfur or iron compounds. The rest is elaboration. There are arthropods and vertebrate animals, insects, crustaceans, arachnids, etc. There are microscopic single-cell organisms, mostly nonphotosynthetic (blue-green algae, for instance) and actinomycetes.

There are fungi that are nonphotosynthetic, neither plant nor animals. They live as long-chain hyphae called mold or mushrooms. Some fungi called yeasts are single cells. There are sprophytic fungi that eat and spew out organic matter. Mycorrhizal fungi create associations with plant roots, taking energy from the plant itself and supplying nutrients to the plant. There are grazers, protozoa, nematodes and microarthropods that feed on bacteria and fungi. And there are microorganisms that transport nutrients into the hair roots of plants once they have achieved a size the microscopic hair can accommodate.

These few notes should be kept in mind as we move deftly from soils to solution botany, keeping in mind that being ocean grown does not excuse the plant from the convenience of soil or its syn-power.

This disparity between land and sea, and the disparity between nutrients left in the land and the abundance present in the ocean presented and affirmed Murray's vision as a scientist.

Was blue-green algae a potential? He tested a variety in hopes of fulfilling his vision. If grown entirely in ocean water, how would mineral uptake be separated from inorganic minerals clinging to the plant? That project, recalls Ed Heine, went down in flames.

In the May 1982 edition of *Acres U.S.A.,* Heine revisited his 30 years with Maynard Murray. Since most of Murray's papers and data vanished with his estate, we are required to consult Heine's assessment of the brief experiment.

"The only factor we have had that has not changed grotesquely has been the ocean," wrote Heine. "Ocean water contains a complete spectrum of elements in a soluble form, whereas soil and fresh water do not. Plants in the ocean can select any and all of the elements they need to grow." This stock fact serves up a question that hadn't been answered because it hadn't been asked. Why not recycle the sea itself back to the land from which its rich nutrients evolved in the first place? Heine's meticulous records say that when the elements present in ocean solids are returned to the soil, plants are able to use what they need, and only what they need. If returned to the land, these elements retain a life-sustaining relationship with each other. This enables ocean-fed corn to uptake 48 elements, wheat and oats 36, soybeans, apples, peaches and other fruit over 30 elements. Most vegetables, including peas, potatoes, tomatoes, carrots and lettuce use more than 25 elements. Records reveal a 12 to 50 percent increase in vitamins, minerals and crop yields, all under field conditions. In tomatoes, vitamin C has been shown to increase up to 24 percent, for carrots, vitamin A up to 40 percent, citrus crops – oranges and grapefruit – vitamin C increased up to 30 percent. The classic 16 minerals and enzymes exhibited an increase of 216 percent. At the same time, the acid level clocked in with a drop of 78 percent.

This balanced mineral availability permits plant life to form and produce proteins, carbohydrates, vitamins, enzymes and the "unknown factors" still to be discovered.

Heine reports with pardonable enthusiasm how various disease agents were sprayed on plants grown with ocean solids and how they resisted the disease, while at the same time untreated plants simply folded and often died. Science seems to deal with the phenomenon by manipulating genes so plants can withstand disease, a procedure that caused William A. Albrecht to comment, "Ah, but can they genetically engineer a plant to withstand starvation?"

Heine concluded, "Millions of unseen forms of life within the soil can do their job more efficiently because they have more to work with. There is more nitrogen fixation in the ocean than on land. This is due to the fact that ocean elements are ideal media for azotobacter nitrogen fixing."

With semiretirement in the offing, the move to large-scale, open-air hydroponics came quite naturally. It was an era during which spin masters proclaimed the end of land agriculture and fortunes to be made under roof with a magic formula and expensive greenhouses heated with natural gas.

Often, good horrible examples contain the germ of an idea. The inherent deficits alleviating the use of difficult-to-store sea solids have been described. But in a Florida location, Maynard Murray could reconstitute seawater and use the water directly on plants in gravel beds. The greenhouse industry had already telegraphed its failure.

Hydroponics was once characterized as the art of growing plant life in a sterile medium using N, P and K nutrient solution. It sounds simple and indeed was simplistic. People identified with the idea because almost every child has accomplished the basics of growing a sweet potato in a glass of water. The experiment proved that plants would grow in the absence of soil, organic matter or microorganisms. Enlarged, this child's experiment can be said to prove that hydroponic growth can occur with inorganic chemicals only. To a large degree, the absence of microorganisms appears to stay free from pathogens, albeit not without sterilization.

As usual, the inventors have come to the rescue. Some 50 years ago, it was established via disciplined studies that pH could be adjusted with sulfuric acid. It was decreed then that the pH environment needed to be 5.5 to 7, depending on the crop, and pH was seen as a limiting factor, along with light, temperature and air. Tomatoes asked for pH 5.5, according to this model. Lettuce wanted pH 7. It was recognized that tomatoes had a problem with uptaking nutrients. The product delivered by this scheme was cosmetically beautiful, but it proved tasteless. It made for superb four-color ad produce and helped borrowers write their prospectuses for loan applications. Usually the equipment fabricator made most of the money.

Greenhouse production is labor intensive. Almost always, hydroponic tomatoes are grown under roof. Plants are well cared for. Each plant is pruned and treated. At the time Maynard Murray moved to Florida largely for the purpose of transferring his experimental knowledge to gravel beds and seawater feeders, the roofed counterpart of set fertilizer hydroponics was proving to be a banker's nightmare. It turned out to be an inefficient system. A short time later the state of Connecticut issued an energy report on various industries, and agriculture was cited. The purpose of the report was to name industries that would have to cut back on energy use in case of a crisis. Energy-wasting industries were to feel the axe. This 131-page report had numbers that were not without interest. To raise one pound of lettuce in the field, for example, took 3,460 BTUs. In a greenhouse in soil, the useage was 29,338 BTUs. When the economists factored in fertilization, heating, chemicals, etc., to produce that same pound of lettuce, the standard hydroponic greenhouse required 65,773 BTUs.

Most of the hydroponic units set up in the late 1970s and early 1980s tasted insolvency within a year. In most cases the entrepreneurs were dilettantes with no background in horticulture or farming. Luther Thomas, a scientist who used greenhouse beds for controlled experiments, told of a North Carolina case to a rapt Acres U.S.A. audience in 1978. A textile manufacturer bought a chemical unit. The firm had never been in agriculture, Thomas said, except as a buyer of cotton. He found out soon enough that hand labor, hand pollination, was staggering. In this case he used heat from the textile operation via pipes to the greenhouse. He packaged hydroponic tomatoes for mail order. Unfortunately, his product had poor keeping quality. They arrived either spoiled or dehydrated and were without taste. Taste becomes a victim in such operations because it is sacrificed for yield. So-called beefsteak tomatoes deliver tonnage sans taste. Shipped in foam material, heat from the vegetable accounted for spoilage. Usually it's such naiveté that brings on bankruptcy.

There was a risk to hydroponics when Maynard Murray moved to Fort Meyers, Florida. He knew the deficits of the system. He expected Florida to exempt him from the ruinous energy requirement. His beds were to be in the open. These alkali N, P and K solutions would be replaced with ocean water, for all practical purposes. Murray was probably aware of the unnecessary practices commercial houses were using to become economical.

The business of using battery acid to adjust pH was a scandal. Use of an organic acid seemed indicated, but ignorance is ignorance wherever it is found. Sometimes counterparts explained that by using battery acid, lead and mercury were being inoculated into the vegetables and then dumped into water tables. Of all non-radioactive toxins on planet Earth, mercury is the most lethal.

Evil Molecule = Slow Death

It takes over the four oxygen sites on the red blood cell. If all are taken over, death follows promptly. That, in a nutshell, is the story of mercury. All else is merely elaboration. It is possible for three and seven-eighths of the sites to be taken over and still have life.

Mercury, whatever its source, kills the oxygen off its normally occupied sites. This is a way of saying the oxygen cannot compete with this non-radioactive toxin. There is a connection between mercury and renal disease, which may explain why dialysis clinics are springing up like shacks in a gold field. Mercury has a stubborn cling capacity. It defies testing and gives up the sites it occupies with pig-headed reluctance. Cellulose, a plant enzyme, will pull mercury out of the body and reduce the burden greatly, but without a kidney function, only redistribution takes place.

Students of the subject not affiliated with government-approved clinics claim that renal failure is caused by a mercury overload and parasites. Mercury has to be chelated out, albeit not with EDTA or IV DNPS, according to this line of thinking, or the IV push.

Whatever so-called settled science turns out to be, finding the source of mercury poisoning rates first attention. Its transport to the brain forces a read of history, the careless growth of technology, and the needless expansion of its use in plastics, building materials and medicines stands as a monument to the stupidity of man. Its use to cure syphilis is legend, and so is the danger of the cure. It has become and remains the base poison on the planet. Once mercury becomes involved, a dementia takes over. Poisons become medicine and medicine becomes poison.

DuPont rode to fame as a munitions maker. Modern war without mercury is not possible. Nor is good money without gold, and gold invites mercury to its historical extraction process. The rule of thumb among the sourdoughs was that to get one ounce of gold, it took four ounces of mercury. The curse of gold – Robert Service style, the debilitation of the miner – has to do with metabolic corruption, the mercury vapor being implicated.

Like the dental folly, gold is an amalgam product. Mercury releases gold trapped in ore. Early on, alchemists tried to turn base metals into gold, the universal currency. Alexander Pope best described the high and mighty leaders who could afford mercury as medicine:

See the blind beggar dance, the cripple sing,
The sot a hero, lunatic a king.

Then as now, doctors killed their patients.

When DuPont decided everything could be made out of a test tube, it was the end of organic man. Hemp gas, hemp oil, hemp medicine were proscribed. Cannabis-based medicine became anathema. The wipe-out enabled DuPont, Hearst, Getty and their lackeys to reorganize society. Marijuana became tied to parts of society deemed alien, and the lie called coal-tar-derivative drugs became the "Matrix" slavery of modernity.

When DuPont patented biodegradable plastics in 1937 — stronger than carbon fiber — mercury was used in the process. Mercury became a constituent part of paper manufacturing, concrete products, automobile motors, and, not least, mercury came to medicine.

The new medicines do not work with nature. They short-circuit and destroy the components of the body while treating one specific thing. All are toxic packages.

The unfolding consequence has been the outlawing of holistic medicine. Cannabis, extracts, tinctures, all have been relegated to the underworld. Food, fuel, fiber, clothing and crop pests were no longer a factor, reformulation being the new deity, but the throne of this new deity had to be made of mercury.

Example: A new car is almost 90 percent plastic inside. Leave the windows rolled up for two or three days in hot weather, and a haze will coat the windows. That's the chemistry of mercury escaping.

Mercury stops fungus. It stops it in well over 4,000 products. Preparation H, spermicides, vaginal lubricants, all survive in the market because they comply with the modern make-it-to-the-door test. If you make it to the door after using the product, you haven't been hurt.

Men who build medicines build labs. The weapons of mass destruction are afloat on the planet everywhere. One of the greatest weapons has been put in the mouths of people with the blessings of scientists and dentists who are not much more advanced in real knowledge than shamans in the Stone Age. Resultant neurological damage is evident, especially in those with too much ready access to fine hospitals, legislative chambers and imperial stations of government.

The lethal passenger rides along in the red blood cells. The body sees such cells as non-cells and destroys them.

The 1974 dumping law must have concluded that plastics were too toxic to be dumped. Therefore, the judgment was made to use the stuff as fertilizer, thereby feeding it back to people, much as fluoride is fed to people in the water supply.

Among 1.25 billion pounds of pesticides for crop rescue, all the industrial waste fertilizers, and all the heavy metal wastes lives the hapless human being. The only caveat seems to be that metaphor in Genesis about tasting the forbidden fruit: the forbidden fruit may have been the tomato, not the apple!

So far, the only remedy from the leaders has been to ship industry out to poverty pockets of the world.

For 160 years dentists were asked to sign a pledge never to use amalgam. When a majority voted for the code, the new leaders formed the American Dental Association. The decision is understandable. The proponents of mercury fillings already had poisoning and were exhibiting a loss of mental acuity. Gold fever was ever triumphant.

It can be calculated that the people in power have a higher mercury load than the nonindustrial worker, perhaps as high as a concrete finisher. The Pentagon drills and fills. Gulf War soldiers have an average of 12 fillings. The shots they receive often exceed two dozen.

There may be evil empires. Certainly we have weapons of mass destruction, even if none are in open view. Above all, we have mercury, some 20 million tons a year, according to best estimates. The evil molecule equals slow death.

The consequence of using battery acids to adjust pH in hydroponics cannot be exaggerated. It invites digression in this book on ocean energy agriculture because pollution near the shore has become a necessary factor, the spill of toxins having made an area the size of Rhode Island at the mouth of the Mississippi a dead zone and coral reefs in the Caribbean a target. As we mentioned earlier, of all non-radioactive toxins on planet Earth, mercury is the worst.

It takes every enzyme ever molded by nature to combat the hazards of modernity, and every enzyme requires one of the trace elements found in ocean water.

Of all the hydoponic units sold in those days, probably 90 percent failed. Most of this record was a consequence of fraudulent promotion, outrageous promises and fuzzy bookkeeping regarding the real economics involved. There is no clear inventory of manufacturers who passed from the scene rather promptly. A few have stayed on.

The penny stock gold mine scam has been compared to the greenhouse hussle. A fellow takes some moose pasture and dubs it a gold mine. Shares are sold. The backlash is that nobody makes money with penny stocks. This is false, of course. Lots of money is made with penny stocks – not in buying them, but in selling them.

Much the same is true of the hydoponic greenhouse, regardless of what the four-color brochures say. And they say plenty. "Crop yields are prolific. Expect 26-by-128-foot magic gardens per acre to produce 240,000 pounds of tomatoes annually. It would take 182 acres of average tomato land in the United States to produce that amount. With a new lettuce-growing technique, a given area will grow 30 times the yield of the land. With cucumbers, the yield is 50 times the yield of land."

One firm produced a claim in a CPA prospectus: "In 365 days of growing, their average tomato plant produced 42 pounds." The comparison is made between soil and hydroponics, not between greenhouse and greenhouse, and certainly not between ocean solids-grown and chemical hydroponics.

In fact a good, balanced soil free of rescue chemicals will produce crops equal to hydroponics. Real hauls on production are *Guiness Book of World Records* stuff, as Charles H. Wilber illustrated in *How to Grow World Record Tomatoes*. His yield was *342 pounds* per plant.

Commercial hydroponics dumps its spent solutions once a week. Some units flush up to 150,000 gallons of this liquid, with its chemicals, pesticides and organic compounds. In some cases, it enters the groundwater system for mischievious transport, God knows where.

The Lorelei's song in any country is just the same: "Growing conditions are ideal. There is no worry from snow, hail, wind, rain, drought, soil diseases or harmful pesticides." In fact, spraying in the greenhouse is mandatory because pests multiply as if to obey the biblical injunction to increase and multiply. Once pests get a toehold, ruin is imminent. In a hydoponic greenhouse, where one plant wastes away, they all waste away. Exit is rapid. Thus the penchant for sterilization, pesticide use, fungicides! Sprays get into the gravel bed. When the water flush arrives, whole molecules of pesticides enter the root system and occupy the plant system locally. These microorganisms described earlier in this chapter are not present in a gravel bed. Therefore, they cannot break down the chemicals of organic synthesis that are part of N, P and K hydroponics.

That systemics in the plant are illegal does not offend the producers of those cosmetic tomatoes. To survive nature's revulsion, hydoponic plants are medicated greatly. They have little or no resistance to disease. They seem immune to plant stock. They can be recorded, trimmed, reinserted into the gravel bed, all without shock. They can be moved from truck to truck with impunity. This is full-grown plants with flowers and fruit.

Transplant any plants — rosebushes, trees, tomatoes — into soil, and there has to be a period of adjustment, usually die-back. Transplant hydoponic tomatoes into the wet organic soil, and presto, the plant is gone.

Maynard Murray sought to make an end run around these deficits with ocean solids dissolved in water. Biologically active solutions did not debilitate plants or require genetic engineering. Using ocean solids, he did not get the results that ocean water delivered.

What was it in ocean water that was erased by evaporation on the shore? People who raise saltwater fish supplied some part of the answers. It is the balance Murray discovered during his trips around the world. Saltwater fish require bacteria or they go belly up. Always, ocean minerals asserted themselves. They keyed enzymes. These permitted transports into the few recedes of plant life. Ocean water was alive.

Claude Carson of Intertech made it possible for researcher Luther Thomas to test enzymes together with seawater in a hydoponic and soil test. The character of the hydroponic plant changed. It was no longer yellowish. The leaves turned blue-green. There was no shortage of nitrogen. Flowers came on rapidly.

As pointed out earlier, men cannot fabricate seawater. That missing cobalt, tin, selenium, cadmium, these exist only in the bodies of microorganisms. The error of stabilizing pH with a harsh acid is thus revealed.

Bacteria produce various enzymes. Nitrifying bacteria not only supply nitrogen, they also release growth stimulants. Bacteria also secrete – that is why citric acid ruins the laboratory stuff, hydrochloric acid. Citric acid is a by-product of bacterial action. Is it possible that those few ocean bacteria in tons of species and other creatures represent early life? According to *Magnificent Microbes,* yes! The supreme function of these proliferating bacteria is to produce hormones, growth stimulants and regulators. They fight off pathogens.

That is why Murray found he could take ocean water plants and transfer them to the soil with impunity.

Now we know why seaweed has to be broken down, otherwise bacterial action will cause the contents to explode. Otherwise, seaweed won't package. Ocean bacteria are the sturdiest life forms in the world. They digest ships and volcanic ash. They eat up refuse, and yet it does not seem to disturb the mix. Metals become molecules and elements and traces under its assault. Likely, the bacteria we give laboratory names are merely imitators of the creatures that once covered the land from Everest on down to the sea level. Left to fend for itself on the land, bacteria adapted to the environment, its supreme purpose.

As nutrients migrate to the sea, they come home to the nutritional center of gravity.

Review Maynard Murray's life work, and we have to concede that he knew hydroponics could never fly as long as it relied on petroleum-based nutrients and as long as fossil fuel energy is used to heat greenhouses in harsh climates.

As a physician, Murray knew that almost all *in vitro* laboratory experiments require agar, which is really just the gelatin of seaweed. The eye, ear, nose and throat medical practice he had during his long career made the ocean connection for him every time. Seaweed extract is the most favored for bacteria. Bacteria require inorganic material, for which reasons agar has the most perfect balance for research. Bacteria are basically animals or plants. They live in water, since all cells must have water.

It was an awesome legacy Maynard Murray passed on to Don Jansen when he asked the professor-farmer, "Why don't you buy my farm?"

Murray was well past any retirement age, albeit still carrying the professional brand of a hospital administrator. He could not have known that he had only one year to live. And Jansen did not know that he had only one year to benefit from a one-on-one tutelage. During that year Don Jansen not only operated the farm, he was tutored like a college freshman on everything from molecules to plagues. "There are anerobic bacteria in our tissues we've never heard of," the physician said. "As the oxygen level is reduced, as we overload with cooked food, we diminish the staying power we have."

"He understood life," Jansen summarized. And he made every effort to confer his knowledge on his protégé. His medical practice was lucrative, but Murray wanted to help all mankind, not just movie stars. Mere money was the least of his desires.

Anaerobes have a hard time of it in the ocean because of the hydrogen peroxide present in seawater. Is it too much to suggest that water from the sea forms the microbial life in a plant much as it governs the plant's hormone and enzyme systems? Having completed his soil experiments all over the country, Murray came to believe that most soils were so toxic it would take eons to clean them. The move to hydroponics was the same chemical pollution under shelter, except in this environment the cleanup store was a *fait acompli*. In river rock, experiments could continue, while at the same time produce moved to growing shelters in the Fort Meyers area.

As he took fruit from his plants in the newly acquired hydroponic beds, Don Jansen said to himself, "Plants do not need soil, not if they get the full bag of nutrients." The soil will always have to change inorganic minerals to organic food for root uptake. Intake through the stomata of leaves was a consideration when seawater was sprayed, but this reveals an area still to be researched.

These carbon atoms cold be joined to an inorganic element in the hydroponic bed — that was a truth the farm presented with its technology. Don Jansen exhaled when he retold the lessons he'd learned, one being that the very same carbon atom exhaled via human respiration is used again by plants.

The beds continued to produce. Jansen could give outlandish guarantees about the shelf life of his tomatoes, and about quality and taste that no one growing with fossil fuel, including oats, could deplete.

As an experiment, Jansen grew plants in tubes — really, the second grader's experiment with the sweet potato all over again. All fruits and vegetables have to pick up their maximum of trace nutrients to be really healthy.

It has been estimated by ecologists that the weight of microorganisms in the world is 25 times the weight of all living animals. The authority for that statement is Bernard Dixon, writing in *The Magnificent Microbes*.

The difference between living and non-living tissue, as Murray put it, is that "there can be no life without a transfer of electricity. Each cell is a little battery. It is capable of and does indeed put out a current. If it is unable to put out a current, the cell is dead and can never return to living tissue. Anything living alters its environment for its benefit in order that it may live and reproduce."

Life is contained in the cell. It is confined to a definite value quite unlike inorganic things.

8

Sea Energy Agriculture

"That little book changed my life!" That's Don Jansen speaking, relating the events that brought him and his father to a chelation clinic in Indiana to see a holistic doctor. Jansen had taken his father to no less than two heart specialists in Denver, where the older man had endured two weeks of hospitalization, whereupon his son learned that not a single test had been performed. Jansen rejected out of hand an offer to perform heart surgery, bundled his father into a car, and drove straight through to the only chelation doctor he knew about. While they waited in the clinic, the elder Jansen passed out. "My dad began passing out on the farm. He'd black out. I was frantic. He had all the knowledge and I was sure he'd die." A staffer called the doctor in alarm. The chelation doctor barely paused after looking at his new patient's chart. "Give him raisins and oatmeal cookies," he told the clinic staffer. It turned out the elder Jansen didn't have a heart problem at all – he was hypoglycemic. "There is absolutely nothing wrong with his heart."

If Don Jansen had caved in at the Denver hospitals, the elder Jansen would have been on the cutting board twice, once for a valve, once for bypass. "That's when I started learning about health. That day in Indiana in that doctor's office, I wondered,

'What in the world is going on?'" Jansen later recalled. He's been studying ever since.

The year was 1980. The little book that changed his life was *Sea Energy Agriculture,* by Maynard Murray and Tom Valentine. After reading about ocean solids and Maynard Murray, Don Jansen concluded that his own three master's degrees and two bachelor's degrees made no difference.

While the elder Jansen was taking chelation, Don joined him for six treatments just to experience the procedure. A man from Arizona who owned five restaurants sat beside Don Jansen. During one of the treatments, Don offered him some buffalo salami. It was a very mild product, with hardly any fat. The Arizona gentleman loved it and wanted the salami furnished to his restaurants. This wasn't possible, but the two men kept talking. The man said he had a book Don should read, and promised to send it. When Jansen got home, the book was there. He read it with mounting enthusiasm. "I said to myself, this man is right on," Jansen recalled. Here was an answer to fertilizer problems. Right there, Jansen decided to put ocean solids on his father's land.

He'd left the family farm in Nebraska while a teenager, trading work for an education. A master's degree at Princeton and a Ph.D. at Ohio led to a tenured position at Indiana State University. Then tragedy struck. A brother who was operating the far-flung range and row crop farm took his own life. He had been ill beyond endurance, hardly functioning as a consequence of the so-called conventional farming that county agents and their advisors had drummed into postwar agriculture. Jansen's father was in no condition to run the farm because of recurrent tumors and other anomalies related to chemical farming.

The farming community in the area was Mennonite, but this was a progressive bunch of Nebraska Mennonite High Plains farmers. Every spray boom, farm chemical, seed "improvement," anhydrous tank, tractor and rig known to the advertising staff of *Farm Journal* filled their sheds. Buffalo and bovine grazed pasture land that was assembled during the parity years, 1942 to 1952, and enlarged some more as the Committee for Economic Development's Adaptive Program for Agriculture became public policy and the so-called surplus farmers left agriculture. "Get big or get out" was the USDA's working doctrine in those days. As size and efficiency became the norm, Don's brother figured he was to be one of the anointed ones, a survivor when lesser men failed. After he got back from the Korean War, the elder Jansen and his son grew the farm from three-quarters of a section to 15,000 acres of pasture and cropland.

"The more acres, the more profit!" That had become *Farm Journal* holy writ. Seldom mentioned was the incubus of expenses that grew the figures on the income statement. Payments weren't instantly lethal, and yet they couldn't be wiped away.

The Jansens enlarged their farm to encompass 5,000 acres of wheat, 600 head of cattle and 50 head of buffalo. The farm was massive, 30 miles across. It took special radios to communicate.

Had they been able to speak, the buffalo could have pointed the way. As a nutritionist, the buffalo is smarter than any Ph.D. Throw a bale of alfalfa hay to this grass-eating herbivore and the buffalo will reject it. Without sign or suggestion, the herd immediately found the grass that had been treated with ocean solids and clipped it like a lawn mower. They rested on those treated acres and wallowed on them. Buffalo love their grass, and they never seem to contract diseases. Over the course of 20 years, the Jansen buffalo never exhibited tumors. A calf might become a victim of the herd moving too fast, but Bang's, Johne's and the legion of diseases common to cattle simply do not affect buffalo.

From the point of view of cycles and history, it was exactly the wrong time to expand, contract debt, and substitute machinery for grass. "My brother tried everything new. He was going to 'do it right' – get big and make it work."

"I'd come home from the university to help out in summer. When I first noticed all those products in the shed, I couldn't believe it. This is American agriculture, high tech?"

"I almost killed a hired man, and that woke me up. My big 70-foot boom strained a hose and broke it. Gas entered the pickup cab next to the machine. He couldn't get out of the pickup. He inhaled the anhydrous. He almost died in there."

Don resigned his university post in 1978. He returned to the high plains to run the ranch, probably to liquidate it if public policy continued to make it impossible to function at a wage delivery profit.

It was then that the tragedy of the moment eluded instant comprehension. His father had been fighting debilitating tumors for years. Doctors identified his heart as the problem. They speculated and guessed and ran up bills faster than the national debt. And still there was no relief.

Don Jansen made a pragmatic observation. The sheds were full of every poison imaginable. This was a high-tech farm as decreed by the universities and the authors of double-track advertisements in *Successful Farming*, as well as a payback full of freebies that filled the mailbox each week. There were skulls and crossbones everywhere. Jansen cut off the supply of poisons, anhydrous, nozzles, plastic tanks and mixers, everything. He didn't know what he was going to do, but he wasn't going to indulge in the health-shattering technology that made life intolerable for his father.

It was this sequence of events that brought Don Jansen and his father to the chelation clinic where he heard about Maynard Murray's initial experiments with ocean solids. Murray had retreated to Fort Myers, Florida, trading his ear, nose and throat practice in Chicago for a general practitioner position at the Sunland Training Center, a Florida state institution for mentally retarded children. He was drawn to Florida because that was where a totally hydroponic form of seaponics was emerging.

When Jansen connected with Murray, he was told that sea solids were available there, "there" being the Baja Peninsula of Mexico. It would take several weeks to order up a semi-load.

The load arrived in due time. Having returned to the farm to help his family, Don Jansen was viewed as an egghead, a school man without the foggiest notion about much of anything having to do with crop production and animal husbandry. This opinion seemed confirmed the morning a petite semi driver called for directions so she could deliver a load of salt. The road was muddy and there were miles of it, with only a small patch of gravel now and then.

She didn't mind. She knew how to handle a rig, she said, adding that she'd go slow. As she snailed her way down the road, farmers whose land had been rained out got into their pickups and started following the semi. Just what is that loco professor up to? By the time the rig got to the Jansen farm, half a dozen pickups were following it. This was fortuitous, since there was no way a tiny woman could handle sacks of sea solids, even on pallets.

When it was revealed that the load was salt, the scoffers were no longer polite. Don was going to fertilize his land with salt. They chuckled while unloading the rig. Each sack had to be hand-carried to a shed. Nonstop chatter punctuated the job for at least two hours. These old Mennonites hadn't enjoyed such loading in years. Each sack removed from the truck inspired a new joke. The egghead was going to put salt on his land. If any of the men had known about Rome salting Carthage, they would have split their sides in bellowing amusement. "I didn't care," Jansen later recalled. "I had Murray's book, and I knew I was right."

After they left, still laughing fit to beat the band, and the semi vanished back down the road, Jansen got out a spreader and put ocean solids on his buffalo pasture. Rain came and pushed the solids deep into the turf. "Those buffalo loved me," Jansen recalled. "They came out of the corners. Buffalo are cantankerous animals that like their privacy." Usually, they back away from the road and resent tourists taking pictures, quite unlike Longhorn cattle, who seem to enjoy their picturesque role. Jansen's buffalo lost their shyness as they munched happily on the new grass.

The Murray connection led Don Jansen to make a decision that turned out to be a grave blunder. He purchased Murray's hydroponic farm in Florida and assumed he could run it and the Nebraska operation at the same time, working one in the winter, the other in the summer, year in and year out.

The effort lasted one season. "You just can't be away," Jansen conceded. The last year he kept the Nebraska property, he ran out of starter fertilizer, basically nitrogen and phosphorus. This material was spread from a 500-gallon tank behind the drill, covering about 60 feet at a pass. Using two units, he would do about 5,000 acres in two weeks. In the last half-section, he ran out of starter fertilizer. It was a feet-to-the fire situation. Jansen needed to get to Florida. He didn't want to initiate tanks again and end up flushing them out, thus eating up more time when he ought to be on his way south. He considered that he could fill the tank with water. Then when he came to line, he could add in Maynard Murray's crystals. Why not add them to the water and recapture the essence of the ocean? "I got a red coffee can, filled it, then dumped it in while the hired men weren't looking; I was that embarrassed. If it worked, fine, if not, the water will help the seeds." Jansen said. Two units were used, two coffee cans and 1,000 gallons of water for a half-section. "That wheat was the best wheat I ever grew. It came out of the soil. It covered the hill like hair on a dog's back. Everyone who drove by saw it. "What did Jansen put on that field?" That became more than a rhetorical question. It came to haunt a decision already made.

Jansen and his father worked the farm-ranch dark to dark, and still they couldn't make much of a living, earning about $10,000 off enough capitalization to backbone a factory. Preparing to liquidate the operation, he'd already sold off much of the equipment.

Jansen may or may not have realized it, but the decision was made for him when then-Secretary of Agriculture Ezra Taft Benson told the press that the government was through molly-coddling farmers.

Wheat has an economic law of gravity. Importing 1 percent relative to domestic production makes the price fall to the world trade, usually a price kept high enough for select domestic operations to thrive. There were other obscurities that confronted Jansen. Early in the Roosevelt administration, laws were passed under the "taking" provision of the constitution. The right to farm was trapped between imports on the one hand and government rule on the other. For funny money, farmers unknowingly sold their souls for a mess of pottage. By taking the farms, the government presumed to assure parity for basic storable commodities from 82 percent of the harvested acres. These commodities in turn became the flywheel and balance for red meat, dairy, fruits and vegetables, and all the specialties. The promise of full parity was kept via a loan mechanism between 1942 and 1952, exactly as promised by Franklin D. Roosevelt in the Stabilization Act of 1942. The Steagall Amendment Legislation annihilated that promise. By the time Jansen made his second sale, millions had been driven from the farms, and 2,400 were leaving each week, year after year.

At the second sale farmers came up to Jansen, always when they could speak privately. "What did you put on that section? What fertilizer did you use?" Few if any believed that only a couple of cans full of ocean crystals were used. Usually they'd shake their heads and walk away mumbling, "He's never serious, won't tell you the truth." The fact is, they couldn't handle the truth.

"It's always that way," Jansen said. "My brother wouldn't have believed me."

How could the average farmer believe what he was being told? Nationwide, some 300 farm papers and newsletters told them that toxic chemistry was high science, and here was a professor selling off spray rigs, tanks and cans decorated with skulls and crossbones, all very valuable stuff, and substituting ocean crystals. That just wasn't believable. Jansen's brother had died depending on the very alchemy that killed him.

Jansen sold out before the scoffers all lost their land exactly as choreographed by agricultural advisors, who made it their business to annihilate their clientele. As it happened, the land angel was with Don Jansen. A new arrival to the general area presented himself at the front door. He wanted land. The stranger was a Mennonite who had moved down from Canada. Apparently the gentleman had downsized his debt by selling Canadian land well above the Nebraska price and now stood ready to buy out Jansen for twice what Jansen's father had paid for it. The debts were wiped out, and the Florida hydroponic farm beckoned. The newcomer bought land, the Jansen shop, some equipment – everything went.

The transition from the high plains took about 25 years, with an ascent into higher education, tenured comfort and finally an epiphany of sorts that caused Don Jansen to discover ocean solids and, even more important, health. The pursuit of education lasted some 14 years, but it really didn't kick open the door.

"My concept of life changed," Jansen said, even though he was set for life. Suddenly everything orbited around at least 90 elements held in suspension in ocean water. The elements seemed to tell the world that mankind was starving much like a blind horse knee-deep in oats. It was the health profile that brought life to terms with reality. Once the farm was gone, Don Jansen immersed himself in the subject of health, especially in the effects of hydrogen peroxide as contained in ocean water.

What was it in Maynard Murray's little presentation that took command of Don Jansen's mind and determined his future? The experiments were powerful enough. They broke the professor-farmer away from the conventional N, P and K ignorance, the false promises that Sir Albert Howard said had swept the republics of learning. There were those words of wisdom Maynard Murray delivered to Don Jansen as they visited on the edge of hydroponic beds when the physician drove by on his way home from the hospital.

He pointed out with disarming finality that the ocean all over planet Earth held in suspension 90 elements in proper relationship to each other. He iterated and reiterated the facts of his research in every sea on Earth through the agency of the U.S. Navy. There was incredible uniformity. Always, the relationship between elements was the same. Jansen knew Murray approximately one year, the period between Jansen's purchase of the hydroponic operation and the doctor's death. The full story emerged a nugget at a time.

After graduating with a medical degree, Maynard Murray interned in Boston. It was there that he often ended a full day's work by going down to the docks.

During his time at sea, Maynard Murray found that the ocean contained the disinfectant hydrogen peroxide in its molecules. It shifts this H_2O_2 to rainwater. This compound literally blesses the water and makes it holy. The vastness of this ocean water, some of it seven miles deep, ridicules man's attempts to balance nutrient elements. This explains why river fish develop cancer the way a drunk gets cirrhosis, and the freedom of the ocean saves salt water animals from that fate. The same turtle on land was always smaller that the brother or near relative in ocean water. That a whale capable of swallowing seals and holding them in its first stomach chamber for later digestion relies on important enzymes was news to Jansen. Murray told him about performing necropsies on two whales, one an infant and the other fully grown. The laboratory reported that it was unable to tell the difference between the tissue samples even though the older whale was perhaps 80 to 90 years old.

Here was that elusive fountain of youth, and it was the ocean itself. The key was the oxygen in the water, and the "keyettes" — if such a word can be used — were the trace minerals.

It was Maynard Murray's thesis that soils had become so poisoned as a consequence of commercial technology that it would take eons to clean them up. This led him to consider hydroponics as the ultimate experiment tool. With this method, everything could be grown in river rock.

When Don Jansen bought Murray's hydroponic farm in Fort Meyers, he found that plants could thrive without soil if they have a flush of nutrients twice a day. Soil is handy because it can take inorganic minerals and make them organic for nutrient support, a function ocean water handles with its carbon capability. Maynard Murray had three degrees in chemistry and allied subjects, and he read prolifically. He had a photographic memory. He probably provided more information than Jansen could absorb, but one discussion stood out as the Nebraskan immersed himself in the work at hand. "Every organic substance has to have a carbon atom attached," Murray told him. "This is the supreme function of the plant. It manufactures organics. Man cannot do it. Plants want only inorganic. Humans can take in organic molecules, but this is not the preferred fare. This is counterproductive. Microorganisms have to prepare the inorganic substance for assimilation in the human system. Plants will take that trace and install a carbon atom with it so the human being can intake organic food."

The quotation is approximate, as are all quotations that rely on memory.

Carbon, of course, is disposed of when we exhale, returning to plants. Plants in turn dispose of their oxygen in order to capture the carbon.

During that brief one-year orientation period, Maynard Murray's hydroponic solution for validating the efficiency of ocean water became settled science, as far as Don Jansen was concerned. The question presented itself: Could hydroponic operations with sea solids, even seawater, become economically viable? It would take experience and even university validation to find out.

9

The Ionic Ocean
Water Connection

If one single thought towered above the rest as Ed Heine and Don Jansen and dozens of other farmer investigators ran test plots and logged in results, it was the intelligence that ocean water contained the same elements as human blood. By the end of century 20, this was common knowledge among medical students and practicing doctors alike. Dr. Maynard Murray recited it often, not as an article of faith, but as a scientific fact demonstrated both in the laboratory and via *in vivo* trials with dogs and other test animals, finally with human beings. René Quinton made his experiments a matter of record in 1904 under the title *L'Eau de Mer, Milieu Organique* (Ocean Water, Organic Matrix). It was an astonishing work, one that preceded Phil Callahan in going too far.

The hypothesis that emerged from years of study and laboratory experiments should have earned for him France's highest award. Instead, Quinton faced no better than Mesmer's work with magnetism, which is now being rediscovered.

Quinton held that the life force in the unicellular structure that showed in *Homo sapiens* was notably more than early ocean water. A hypothesis is usually considered an educated guess. In Quinton's case, his conclusions flowed logically from the incontroverble facts.

Born in 1867, he passed from the scene in 1925, leaving a storm of controversy in his wake. Unlike Maynard Murray, who merely picked up medical school intelligence that blood was the same as ocean water, Quinton envisioned ocean water as therapeutic, and not merely fertilizer for plants. His postulates became his credo, each more enduring than brass plates on a monument.

Ocean water had to be pure in order to achieve therapeutic status. It had to be collected in ocean zones free of pollution. Further, it had to be diluted to the concentration of primeval ocean water.

The dilution process, using stainless steel tanks and heat, changes ocean water by stripping it of its life factor by evaporation.

Finally, Quinton found evaporated water actually harmful.

It took this French scientist several years to find his answer, one he made a matter of record via a patent designated as "Plasma de Quinton." His book *L'Eau de Mer, Milieu Organique* dealt with more than aspects of ocean water. It dealt with the origins of life — every form of life, plants included. Those who read French could find a veritable encyclopedia in these pages, although for purposes of this exposition, we have relied on a partial English translation by Peter Margo of Montreal, Quebec. Among the translations available are the dog experiments.

Much like Pottinger with his cat slides on California beach sand, Quinton received stray dogs captured by the authorities. These animals were anesthetized. Slowly, their blood plasma was withdrawn. Ocean water was infused into their bodies. As more and more water was infused, these animals evacuated their own blood plasmas through their kidneys. The maximum ever infused into a dog was his own weight. In short, the animal could and did live on ocean water, at a one-to-one replacement.

The many dog experiments are represented as one in these lines, but actually, a series of dogs not only lived on ocean water, they thrived on it. They were healthier after the procedure than they had been before.

Animal experiments led Quinton to employ ocean water as human therapy. At the time there was a veritable epidemic of babies dying of intestinal infections. Quinton set up a dispensary in Paris. The procedure proved so successful, various physicians set up clinics to help the children. A total of 10 were set up throughout the country. Thousands of babies would have died without the ocean water treatment. The case reports tell of weight gain within days after infusion of the water. All were healed quickly; the water brought about complete reversal of the infection. Quinton's ocean plasma exhibited a therapeutic effect characterized as complete reversal of the disease. It was and still is explained as a change in the body's terrain. Simply stated, pathogens do not thrive in ocean water. Quinton's time was an age in which scientists were still trained to turn base metals into gold and a pile of minerals into ocean water. Quinton collected the solids and sought to duplicate the ocean's therapy using distilled water. It failed.

Ocean water has an organic matrix. It is a proliferation that only nature can produce. Its construction at the hands of laboratory scientists is as elusive as the creation of life in a test tube.

There is puzzlement here. No one would want to take lead, cyanide, mercury and some few other elements into the human body. Yet there are elements that are fingerprinted in ocean water that the natural concentrators, exempt from uptake.

Except in Canada and Brazil, ocean plasma therapy has been proscribed, especially in France, where the government has gone to the trouble of making Quinton's discovery illegal. Some Canadians tell us that they have been drinking diluted ocean water for years, with only healthful effects as a consequence. The usual amount consumed as health therapy is a single ounce each day. The records from yesteryear tell the story, as do practitioners of the ocean plasma art today. While preparing this manuscript, I encountered a student of Quinton's art who took as much as a liter of the isotonic water in his colon with no adverse effect.

We tend to look at ocean water in terms of single elements. We seem obsessed with the idea that we can dissect and reduce nature, put each element in a closed, air-tight compartment. Our puny minds can hardly escape reduction as an operating tool, and yet the mosaic of the whole provides the only true picture.

Plants may not pick up this, that or the other element, and yet no harm results from the exposure.

Readers will have noted fluorine is a constituent part of ocean water. Yet Don Jansen's readouts do not show the element in live plants. Apparently, plants do not pick up what they cannot use. Is there a similar government for human cells and protoplasm?

Chemistry is a valid tool. It is also one that baffles the fools and fools the wise. The organic life-giving dust of the ocean demands its own science, the very science that was discovered, then shelved in the nation of its origin.

Ocean water, declared Quinton, has energies. It is ionic, meaning cells of the body and cells of the plant can use it.

Quinton's voluminous studies first proceeded on the premise that ocean plasma had to be injected. This proved to be a false start. The same effect was achieved, soon enough, by absorption through the colon, in turn allowing for absorption by osmosis directly into the bloodstream.

Except for journalistic license, it is not possible to point out that the subjects of these few case reports of Quinton-style ocean plasma treatments never seem to experience even the common cold. Even if cell biologists are right and viral agents never leave the body, they are made quite helpless by ionic ocean water. Flu shots pale into insignificance when compared to an ounce of ocean plasma, this according to associates in North and South America.

The Gerson therapies have a long record for clearing up cancer. Dr. Max Gerson did not use ocean plasma, but he followed the Frenchman's spiritual lead by giving patients the best available mineral load via juices of several kinds. Those who discovered ocean water during the early part of the last century saw their discovery as a logical extension of lessons learned from the pioneers in juicing. Quinton's successful treatment of cancer patients with his ocean plasma water was certain to invite the hostility of the medical profession. Laws were stamped into the books as if by a thundering herd of horses. Quinton treated tuberculosis, cholera, typhoid, dozens of other anomalies, not least cancer. The many diseases that affect mankind are real and moving tragedy. They are also cash cows for the professional medical practioner.

Isotonic water these days is ocean water diluted two to one, the "one" being the dilution. Psoriasis, edema, hair loss – these are entries of more than mere academic interest.

All cellular life began in the ocean. All cellular life has the same plasma. When cellular life began, the ocean was much more dilute. It was at the concentration that blood plasma is now. The process of diluting ocean water to the consistency calculated for ionic water cannot rely on distilled water. Generally speaking, mineral water from springs is used. Quinton held that distilled water was dead water. This much stated, it must be added that mineral water for dilution must be of low mineral content.

Almost as mysterious as the life of ocean water is the character and life of René Quinton. He was born in 1867 and passed from the scene in 1925. It was an era of discovery and innovation by self-educated people. Edison was largely self-educated. Lincoln, America's secular saint, had only one year of formal education. René Quinton was educated or he was self-educated. His first discipline was that of a writer. He came from a small town in France. His father was a doctor.

Somehow Quinton became interested in biology at the approximate age of 28. He got his on-line training in biology in a first-rate laboratory under the tutelage of a leading biologist. During World War I he became an aeronautical engineer specializing in airplane design.

Suffice to say he became famous as his therapeutic use of ocean water became known. By the time he passed away in 1925, he had become so esteemed by France, the President paid his respects at the funeral. The President of France unveiled a statue of Quinton in his home town. None of this mattered as far as conventional medicine was concerned. Once the drug industry achieved power, Quinton became de-Quintonized. In France it is against the law to sell diluted ocean water for therapeutic purposes. The official French attitude has not stopped Quintonesque procedures in Africa, most of Europe, South America, and even in Canada and the United States. Use of

Quinton therapy has proved especially effective with African children who have become dehydrated. Administration has been given orally or by injection.

The story has been told that when Quinton added white blood cells to isotonic water, the white blood cells stayed completely alive – for weeks. The lesson was clear enough: As long as the body can be kept clean, the cells can be kept alive. Simply stated, infectious organisms cannot live in ocean water. Some of Quinton's patients believe the SARS epidemic could be stamped out quickly using Quinton ocean ionic water therapy. Pneumonia and cholera symptoms that are very similar to SARS have been reported cleared up under the auspices of ionic ocean water therapy.

To a world that thinks of health as coming from a drugstore, the esoteric nature of nature's choice is troubling. Can the ocean really digest tanker spills, sunken ships and rivers spilling their refuse into the sea? The answer seems to be yes and no. It is not possible to take water from the ocean near a shore for agriculture, much less for the highly technical refinement required for human therapy. Quinton's requirements were exact. The water had to be 20 meters off the ocean floor and 40 meters below the surface. The area for sampling had to exhibit special norms, such as eddies, ocean streams, and other anomalies. Indeed, each operation required newly lined tanks. No one should read these lines and conclude that a trip to the beach for a swig of ocean water constituted anything more than ocean's revenge.

The point here is that ocean water retains its molecular balance as a living medium, and therein lies the benefit for crop, animal and mankind.

The conclusions established by this 19th century worker have a validity of maximum interest to farmers, nutritionists, physicians and the people called patients. All organisms can be considered an aquaculture filled with a liquid called the primary ocean. All the cells of the body require this bath. When the quality of the bath changes, cellular nutrition is destroyed, this in plant, animal and mankind. This undermines the defenses of plant, animal and mankind against pathogens. Therefore, to opt for health is to opt for restoration of the liquid envelope in which plants, animals and human beings live.

There are no secrets in nature. There is only a failure to see what we look at.

Brazil's Dr. Francisco Antunes of Instituto Augusta de Pesquisas Bioquímicas in Sao Paulo has been using ionic ocean water therapy for 40 years and now has a voluminous body of research validating his procedures. His career is summarized in *Terapia Ortomolecular Natural,* published in Portuguese. A focal point in his work is the chronic shortage of elements in the food supply. Hidden hunger, the anatomy of salt, and ocean water as Earth's blood — ocean plasma — are some of the tantalizing subjects with an ocean water connection.

It is now estimated that the mixing time for the ocean is about 1,500 years, but it takes a water molecule about 38,000 years to make the round trip from land to ocean and back to land as rain. Therefore, the ocean is mixing much faster than ions are being supplied. That is why the ocean stays essentially constant. Nature's lessons from the ocean, the soil, even the air are not different compartments in a complex structure. They are the same side of the same coin.

10

A Sea Energy Testament

Leaps forward in all phases of science seem always to take place in different channels at the same time. No doubt, Hock discerned the general mechanism of evolution long before Darwin and Wallace. Satellites orbiting planet Earth to make world-wide cell phones possible feted the pages of science magazines in the 1930s, long before Sputnik was flying into space. The concept of insect and weed proliferation was hammered into place even before the beginning of the last century by Russian and European agriculturists, then validated by the University of Missouri's William A. Albrecht and Friends of the Land meeting at Louis Bromfield's Malabar Farm.

Lee McComb, an Illinois farmer, undoubtedly considered Utah's ocean salt and mineral mix a potential source of plant nutrients, nutrients being the best way to control against bacterial, fungal and insect attack. His *Ocean Energy Testament* told the story as an unpublished paper, excerpted below. It recalled that the valley of the Great Salt Lake was once an ocean bed, this perhaps 1,000 million years ago. McComb writes:

> My first introduction to the possibilities of ocean solids as fertilizer was during a visit to Salt Lake City in 1931. After seeing the interesting sights of the

city, we drove to Great Salt Lake to swim. We were amazed by the fact that we didn't sink in the water – we floated around.

I asked one of the local citizens why the water was so salty. He told me they didn't get as much rain as we did in Pennsylvania, but they did get enough to leach out the salts and natural soluble minerals in the desert, wild country and farmland in the large area around the lake. The mineral salts were drained into Salt Lake by streams. He added that Salt Lake had no outlet. The only way water could get out of Salt Lake was by evaporation, and minerals do not evaporate. They just thicken the remaining water. That was why the heavy content of salt in the lake kept us afloat. Then the thought occurred to me – if rains were washing the soluble salts out of the soil into Great Salt Lake, then the lake water should be liquid fertilizer.

In 1944, I finished my service in World War II as a Military Police officer guarding war prisoners in the Pocono Mountains, where the prisoners were paid to help clear 500 acres of farmland for the Starke Farm Company, who had 3,000 acres in their main farm near Morrisville, Pennsylvania. As a result of some suggestions I made to Starkey's farm fore-man, I was hired by Mr. Starkey to be the assistant production manager of his farm.

At this time, the insecticide people were saying that insects were becoming immune to their poisons, so they had to make them stronger. I could see dead insects after we sprayed our vegetable crops, and we did a lot of spraying, so I didn't believe immunity was a factor. However, it started me wondering why we continued to have insect and disease problems.

In my thinking I went back to pioneer days, when they didn't have insecticide or chemical fertilizer or manures. They didn't have enough animals to produce much manure. The pioneers cleared the trees off the land, planted their seed and produced good crops for some time. Then crop production began to come down, and insects began to attack the crops.

This was no big problem to the pioneers. They just cleared more trees off more land and repeated the process. But to me, it meant that the crops were taking minerals out of the soil faster than the natural life in the soil could replace them. The result of these deficiencies was sick plants that attracted insects and disease like a dead animal attracts the crows and buzzards.

In my search for a solution to our worsening insect problem, I remembered reading that our Creator had made man out of the dust of the Earth. The fact that birds, animals and mankind depend upon plant life for their food and that plants depend on the soil for their food convinced me that the report of creation in the Bible was true. This means that man, animals and birds need all of the elements in good soil for good health and insect resistance.

In 1945 we moved to Florida, where we could make compost fertilizer the entire year because of the warm climate. Our compost proved to be quite successful in increasing vitamin and mineral content, as well as improving the flavor of fruits and vegetables.

In 1966 one of our sons, John, had to have a science project for a display his chemistry class was setting up. He asked me if I could suggest an interesting demonstration he could set up. I told him about my experience at Great Salt Lake, adding that the same factors that made Salt Lake salty also made the Atlantic Ocean salty. I suggested that he prove that Atlantic Ocean water was a liquid fertilizer. John said he thought I was crazy but agreed to try it.

We got 36-gallon cans and filled them with washed builder's sand, so the potting soil would have little or no available food value in it. Then we got six varieties of vegetable seeds and planted one variety in each of six rows. Our plan was to use ocean water for both water and minerals to grow the plants. The damp sand germinated the seed quickly. When the plants were 4 or 5 inches high, we put ocean water on the first cross-row section of the six rows. On the second row, we diluted the ocean water by 50 percent with well water. In each succeeding row, the ocean water was diluted by 50 percent with fresh water, until the dilution for the sixth row of plants was 10 percent ocean water and 90 percent tap water.

The first row, which got straight ocean water, did not die as I had expected; neither did it grow. The plants just stayed green. The less ocean water in the mixture, the better the plants grew. The fourth, fifth and sixth rows of plants looked like we had fertilized them, the sixth row being the best.

As a result of this test, I started searching for a source of ocean solids (that is, the complete ocean water dried out to the salt crystals residue). The salt on the market today is a refined salt processed for the water-softening business, and it will test to at least 99 percent sodium chloride. It is not good for growing plants.

The analysis of ocean water is close to 90 percent sodium chloride. That leaves just 10 percent to include all the other elements in earth. The analysis of human blood, I am told, is the same as ocean water. In areas of the world where human blood is not safe for transfusions, ocean water is reportedly used. I am inclined to believe that ocean water would be superior to blood, because ocean water does not have the toxins and risk of spreading disease that human blood might be carrying.

By using ocean solids as fertilizer to produce foods for man, mammals and birds, I believe we have found the source of good health and happiness.

In tests we have run growing wheat in North Carolina, we found the average protein in North Carolina wheat is 12.5 percent. The wheat grown using ocean solids was 16.4 percent protein. The first year's application of 1,000 pounds of ocean solids per acre made the difference. This also tells us how quickly worn-out soil can be restored to fertile soil. No record of production was kept by the farmer who grew the wheat, but he told me that where he used ocean solids, the wheat heads were filled out to the top with plump kernels. On plants where he used his regular fertilizer, there were six to eight places where there should have been a kernel of wheat, but there were only half-developed kernels or no kernel at all. The farmer also stated that the first difference ocean solids showed was that the soil stayed moist much longer after a rain than the other part of the wheat field. This would be a great benefit to the plants during normally dry-weather periods in the growing season.

In 1936, Dr. Maynard Murray became a researcher on a ship, doing research on ocean life all over the world. He wrote a little book entitled *Sea Energy Agriculture*. When Mr. Basil Rossi contacted me about speaking at a conference on ocean energy agriculture, I suggested he contact Dr. Murray as the man who knew more about the subject than any man I knew. Unfortunately, Dr. Murray was having health troubles at the time Mr. Rossi contacted him, and he died May 30, 1983. I would like to tell you about some of the results coming out of research using ocean solids as fertilizer for Dr. Maynard Murray.

Dr. Murray told me that the waters of the ocean hold the perfect balance of those essential elements required as food for the complex cell groups that make up our bodies.

He said of special interest is the fact that the aging process does not appear to occur in the ocean. The cells of a huge adult whale and cells taken from a newly born whale will show no evidence of the chemical changes observed when comparing cells of adult and newborn land mammals.

Science tells us that nearly half of the individual cells in an animal body are replaced during the process of cell division. In man, for example, most of the cells are replaced every 18 months. If the required elements are not supplied by the food ingested as cell division occurs, dilution becomes apparent and continues until these critical elements are nonexistent in the organism. This shortage of essential elements does not occur in the ocean, because in the ocean they are always present and available. By applying complete ocean solids to food-producing soil, our food would supply all the critical elements for cell division in man, animals and plants.

A plant can grow to maturity and yet make dangerous substitutions of elements in its structure, due to the chemical attempts to compensate for an imbalance of the proper elements in the soil. If our cells, in turn, must compensate for the dilution or lack of elements, then they lose their resistance to diseases present or attract diseases and insects.

Food and other crops require an average of 40 elements from the soil. In no case do fertilizers add more than 12 elements, and most commercial fertilizers add only three to six. The single most devastating source of depletion of soil is water (rain) leaching. Growing crops on land is the second great mineral leacher.

The following records of increased production from 1900 to 1960 is a tribute to the productive ingenuity of soil scientists and farmers. It is also an indication of the acceleration of depletion of essential elements in soil fertility. Increase in production for corn yield was 58 percent, wheat 55 percent, rice 54 percent, oats 25 percent, sugar beets 66 percent.

Shortly after moving to Leesburg, Florida, I became interested in growing watermelons. At that time, I was told that you could not grow watermelons on a field more often than once every 10 years, because a crop of melons took so much plant food value out of the soil that the plants would be killed by anthracnose and other wilt diseases before the plants had time to produce melons.

I had been growing watermelons for seven years, finding new land to plant each year, when I decided to go back to a field we had grown a good crop of watermelon on three years previously. Much to my surprise and delight, this test block of melons did not die of wilt diseases, as we were advised would happen. We learned that watermelons are heavy feeds on the elements of the soil. They used up the elements that protected them from attack by wilt disease. Without knowing what elements were needed, we were nevertheless including enough of these elements in our compost to provide the requirements. The plants were therefore not attractive to wilt diseases, and they produced a good crop of watermelons. This eliminated the necessity of finding new ground to grow the next crop.

We thought that the compost-grown melons tasted sweeter than the same melon variety grown the conventional way on chemical fertilizer. In order to get more definite information, we had tests made of samples of three crops. Results were as follows.

In tests by Grey-Scheu Lab in Jacksonville, Florida, compost grown watermelons contained:

- 12.3 percent more vitamin A
- 20 percent more vitamin C
- 50 percent more niacin
- 10 percent more choline
- 41 percent more sugar content
- 63 percent more food value in minerals

In tests by Southern Analytical Lab in Jacksonville, Florida, compost-grown oranges contained:

- 16 to 37 percent more juice per orange
- 36 percent thinner skin
- Up to 78 percent less acid
- Up to 30 percent more vitamin C
- 234 percent more food value in minerals

In tests by Southern Analytical Lab in Jacksonville, Florida, compost-grown cauliflower contained:

- 218 percent more vitamin A
- 40 percent more vitamin B_2
- 183 percent more niacin
- 50 percent more food value in minerals

These tests gave us some idea of how much more nutritional plants can be for human and animal food if given the proper kind of plant food. If we had included ocean solids as source of minerals, I am sure the difference would have been increased. We have never had to spray any of our crops for insect or disease infestation in the last 30 years.

We would like to share with you the following quotes from Tom Valentine and Dr. Maynard Murray's book, *Sea Energy Agriculture*. Dr. Maynard tells us: "The quantitative analysis of elements in human blood has essentially the same profile as the quantitative analysis of elements found in ocean water, including large amounts of sodium chloride. The amount of sodium chloride contained in ocean water will cause many to question its use as a fertilizer, because it is well known that salt has been used throughout history as a way to kill plant life on land. The secret lies in the use of proper quantities of sodium chloride in proper balance with other nutrients; a balance which characterizes ocean water and ocean solids."

The following quotes by Dr. Murray refer to disease control in plants: "In 1940, a plot containing four 12-foot high peach trees located approximately 20 feet from one another was selected to begin the experimental process of determining the effects of our fertilization process and resulting resistance to disease. The first and third peach trees were designated for experimental tests and were treated with 600 cc of ocean water per square foot from the base of the trees to the edge of the foliage to cover the main areas of nutrition. The second and fourth trees were designated the control group and did not receive application. We made the initial application of fertilizer in March, before the trees started to bud. About the first of May, all four trees were sprayed with 'curly leaf' virus, and their peach yield was sharply reduced from the norm. The observation period for the test lasted three years, although spraying with the virus took place only in the first year. The control trees developed 'curly leaf' each year and finally died. The experimental trees retained resistance throughout the three-year test period and provided normal yields each year.

"In the same year, turnips were planted in a plot of soil designated half control and half experimental. The experimental section of the plot was fertilized with 600 cc of ocean water per square foot of soil after the staphylococcus bacteria, associated with 'center rot' in turnips, had been mixed in the soil of the entire plot. When the turnips sprouted and the leaves appeared above the soil line, the leaves of both the control and experimental turnips were sprayed with the same bacteria. All of the experimental turnips grew to normal, healthy turnips without evidence of center rot, while the control turnips contracted staphylococcus center rot and died.

"It was next decided to grow tomatoes hydroponically in a controlled-diet environment, for which the following system was used. Two boxes measuring 100 feet by 3 feet by 8 inches, constructed of cement, were filled with sterilized marbles about three-eights of an inch in size. The tomatoes were planted in tissue paper a foot apart in the hydroponic beds. A nutrient solution stored in a tank was flooded into the beds, drawn back out and returned to the holding tank three times each day. After the tomato plants had sprouted and their root structure adhered to the marbles, the foliage was tied up and pruned to one stalk. The experimental hydroponic bed received a 112-pound ocean solids to 5,000 gallons of water solution mixture, while the control bed used a traditional hydroponic solution. Both beds were flooded three times daily. 'Tobacco mosaic' virus, also lethal to tomato plants, was selected as the exposure disease, and all plants were sprayed. The experimental plants did not contract the disease, while all the control tomato plants died of the 'Tobacco mosaic' virus.

"Hydroponic experiments were conducted in both greenhouses and outdoor settings, and the experiments were later repeated in Fort Myers, Florida, where the disease incidence is extremely high. Experimental tomatoes grown in Florida still remained disease free. In all cases, the taste was superior, and pollination as well as resulting production yields were excellent on the experimental crops. Tomato experiments were conducted in gardens in northern Illinois during 1954 and 1955. Here the experimental plots were fertilized with 2,200 pounds of ocean solids, while the control plots were again administered the traditional fertilizing applications of N, P and K. The control plots indicated heavy blight from fungus; the experimental tomatoes that had been fertilized with ocean solids were blight free."

I offer here Dr. Murray's comments on the nature of life on this planet:

"Life on Earth is divided into three major categories: plant, animal and protista (organisms that display a combination of plant and animal characteristics). Of the roughly 1,250,000 known life species, almost three-fourths are animal, with the remaining one-fourth made up of plants and protista. The ocean, which covers over 72 percent of the Earth's surface, probably contains 90 percent of all life. This is true because of the extent of the ocean surface and the fact that the ocean averages 2.25 miles in depth, with some places measuring seven miles deep. Forms of life can be found throughout the ocean, including the lowest regions, whereas life forms on land are generally found in the upper foot or so of soil. Even land birds and other flying animals must trace their food and life cycles to the soil.

"When looking at ocean life, one is immediately impressed by the fact that in this 71 percent of the planet's surface, there is no cancer, hardening of the arteries or arthritis. A whale carries around tons of fat and lives in an environment of salt solution, which makes one wonder if the medical advice of a low salt and low fat diet is justifiable. Ocean trout do not develop cancer, while a large percentage of freshwater trout over five years of age have cancer of the liver.

"The fact remains that it is very difficult to find any species on land that does not have cancer, while animals living in the ocean are without cancer."

Dr. Murray continues: "It occurred to me that if there were some way to nourish land animals on food that contained all the essential elements, it should make a difference in their resistance to disease. Diet must be the secret. Ocean water contains a complete spectrum of elements, whereas soil and fresh water do not. Plants in the ocean can select any and all the elements they need to grow. In turn, ocean animals feeding on these plants easily obtain their element requirements and thus better disease resistance. If ocean animals can establish their resistance through diet, I feel that land animals can also obtain this resistance if the food which they consume has been grown with all the necessary elements available.

"In the early days, ocean water was trucked into the Midwest and applied to soil considered adequate for growing cereal grains. This, of course, was an expensive method of securing the complete range of elements from the ocean, so we immediately began to look worldwide for natural locations where the ocean water becomes landlocked and total evaporation takes place. The largest deposits were found in Mexico, with others in countries of South America and some in Africa. This complete spectrum of elements from the ocean we have designated ocean solids, and, when used hydroponically, our method is labeled Seaponics.

"We began using these complete ocean solids in growing large quantities of cereal grains for feeding animals. Ocean solids were applied at the rate of 1,000 to 2,000 pounds per acre to half of the fields, while the control half received only the customary fertilizer. At harvest time, corn smut, rust, and other cereal diseases were significantly reduced in the experimental fields. Disease resistance had been fixed in the plants by the use of this complete elemental diet. The next step was to see if the resistance could be transferred from plants to animals."

Dr. Murray reports the following animal experiment: "A first animal experiment was carried out on C3H mice which get spontaneous cancer of the breast. These cancers are most probably due to the so-called Bittner virus. We hoped that by using ocean solids-grown food we could build resistance to the virus or cancer. The C3H mice were divided into two groups. The control group was fed on regular cereal grain, while the experimental group was fed cereal grain raised on ocean solids-treated soil. The results showed that instead of getting spontaneous cancer in 90 percent of the animals, as the control group did, the experimental animal figure dropped to 55 percent. The second generation born to parents fed on ocean solids food produced cancer in only 2 percent of the population.

"We also tried the special feed on transplanted malignancies (sarcoma) in rats. The ocean solids-fed rats showed high resistance to the transplanted sarcoma. A transplant would die off most of the time, but occasionally it would take. After growing awhile, it too would die off. This success occurred in over 90 percent of the animals treated with the ocean solids-grown food.

"Other animals were fed ocean solids-grown food and did well. Dairy cattle were raised in this manner, but no statistics were kept as to production or quality of milk. However, it is interesting to note that if a bundle of corn stalks containing a mixture of ocean solids-grown corn and untreated corn was offered to the cows, they would nuzzle through the bundle and always eat the ocean solids-grown corn first.

"These experiments were continued for three to five years. It was found that the bony structure of cattle and horses fed on ocean solids grain was better. This was also true of chickens. Eggs of experimental chickens would be larger than those of the chickens fed on controlled food. The difference was evident from four to six months of age. These experiments show that changes can be transferred from plants to animals.

"There are several avenues available to us to reclaim and improve our soil and food production. Each choice is dependent upon the amount and condition of the soil and the need and availability of food.

"The first avenue is to use ocean solids on the remaining agricultural land. Ocean solids return to the soil elements lost through overcropping and generally poor cropping techniques. Past experiments indicate that the amount of food produced with ocean solids is generally increased. Past experiments also indicate that ocean solids can improve disease resistance in plants and that these plants can, in turn, increase disease resistance in the animals that consume the plants.

"The second way is to substitute hydroponics for land where the soil is completely incapable of food production. One can produce 10 to 50 times the amount of food per surface area in hydroponics than can be grown on soil. This is especially important in areas that have little or no arable land and large populations to feed. Since nature uses 71 percent of the Earth's surface for hydroponic growth (oceans), it does not seem too far-fetched to consider growing much of our edible food in this manner."

I am convinced we can greatly reduce the need for poison insecticides, if not eliminate the need, by returning the elements washed into the oceans by rain all over the world to fertilize food and fiber crops that are essential to life on Earth.

A warning at this point is critical. Sixty percent of the demand for salt today is for use in water softening. For this use, the sodium chloride content must be close to 99 percent, leaving no room for other elements essential to plants. The ocean solids for use in foods and agricultural plant life should be close to 90 percent sodium chloride, as it is in human blood. This leaves 10 percent to include all of other elements in blood and earth. This means that the amount of nitrogen, phosphate, potassium and other elements can't be more than a few pounds per acre.

At present, we have a supply of agricultural ocean solids and can get more as demand dictates. Otherwise, it is not generally available. There is no shortage of the supply of agricultural ocean solids. The ocean can supply millions of tons per year to satisfy a demand.

However, before the farmer can benefit from nature's gift, the salt must be extracted from the ocean water and made available in a granular form for the farmer to use.

Lee McComb thought the idea of ocean nutrition being used on farm acres was slipping away. In truth, the investigators had not been turned into practitioners, because no entrepreneur had picked up on supplying a trade that simply had not developed. Maynard Murray had discerned the fact that it was necessary to go out some 50 to 60 miles before taking water that could be used. A few more details in this fascinating story are presented in the chapter that follows.

11

The Bottom Line

There is an old Russian saying that what is written with a pen cannot be removed with a hatchet. Maynard Murray's record of nutrient uptake by plants grown in ocean water continued to tantalize botanists long after his hydroponic beds had been closed down and the encumbered land sold by his estate. Don Jansen set up a hydroponic operation in Leigh, Florida, and ran it for five years. It was located on an 11-acre farm built to attract people to buy lots in a development, one of those retirement communities that helped make Florida the lodestone state. The developers had built an airstrip behind the farm. Once they saw the lush garden with its cosmetic crops, prospective customers turned into pushovers. Jansen's beds were located about a block away from the terminal tower, a perfect site for them to function as a fabulous sales tool. Looking back, Jansen concluded that the operation could have been an economic success. Several considerations interdicted this goal.

There was the matter of capitalization. The crop beds required hands-on attention, state-of-the-art equipment and a measure of dedication that was hard to come by in the available labor pool. That, however, wasn't the real shortfall. Public policy by then was well into the business of opening the border for imports. The tomato groves of Homestead, Florida, couldn't compete with south-of-the-border growers who employed workers at poverty wages. Imports captured the produce market from American growers at an alarming rate as the General Agreement on Tariffs and Trade (GATT) laid the groundwork for the North American Free Trade Agreement (NAFTA), and finally the World Trade Organization. Vegetables, seafood and fruits flown in on subsidized airplanes became what economists call a comparative advantage. As long as the fruit or vegetable looked like a fruit or vegetable, the consumer ate the produce, even if it tasted like cardboard and was loaded with pesticides, not nutrients.

Don Jansen ended up selling the property back to the Leigh developers.

"You get ripped off," Jansen revealed. Many a solvent operation runs into a stone wall of liquidity. The enterprise is viable, but cash flow and a profit stream often elude the grasp of even the most diligent small business owner. Jansen went to the Small Business Administration. Yes, they would make a loan, but to do so they required transfer of all patents and technologies to that agency. At least a half-acre of trees must have been sacrificed to supply the paperwork required. Jansen thanked these good people and never went back. Still, there were those haunting numbers, tomatoes picking up 56 nutrients while so-called commercial produce picked up hardly any.

Since 1996 the United States has imported $1.31 worth of goods for every dollar's worth exported. As these lines are set down, the numbers are close to two dollars of imports for every dollar of exports. That means the deficit has to be at least $50 billion a year, and perhaps $100 billion a year for the NAFTA world trade model to continue.

Those who speak in favor of world trade contend that this import-export razzle-dazzle is necessary for world peace. Each nation is supposed to spend what it earns, this so each trader nation can exhibit its comparative advantage. The banana grower grows bananas. Third World countries sell what they can, mainly cheap labor. In turn, the United States sells its high-priced trucks, airplanes and supercomputers. Never explained is how a 25-cent-an-hour worker will be able to buy a pickup truck or a bottle of French wine. All this leaves unanswered the question of just where is the necessary trade expansion going to come from when one and all are tapped out?

That said, it seems the American farm model was a poor one for international trade to follow. The importance of farm products is twofold. Jansen put it this way: "We're not talking about the outside of the plant, we're talking about the inside. Everyone thinks in terms of bins and bushels, pounds, but people are getting so sick, they're starting to think of what's inside."

Selling through established markets, shipping vegetables to Winn-Dixie or Safeway and picking up a check in competition with Mexico and Guatemala, is not a business farmers can enter with savings and expand with earnings.

Don Jansen did not go back to the drawing board immediately. After selling the Leigh operation, he continued his research and self-education on nutrition. A rancher from birth, he virtually abandoned meat protein foods. He wanted live food, food full to the brim with enzymes, food loaded with the nutrients ocean water stored in abundance. One of the investor types who visited the Leigh operation now and then was a gentleman from England, Paul McCrary. He was more than fascinated with Jansen's hydroponic operation. The taste of everything grown was fantastic. More important, McCrary knew how to crunch numbers. He believed in that nutrient pickup, of course, and the time came when he was ready to invest.

The system of mainstream research currently in force makes science the validator. Disciplined work and thorough documentation, not anecdotal evidence, carries the day. McCrary wrote out a $100,000 check, and Jansen found a university willing to do the tests and write up the obligatory paper — in a word, validate what Maynard Murray and Ed Heine and Don Jansen knew all along.

A few restaurants grew vegetables for their customers using ocean solids. One was Lee's Steak House in Florida. The bottom line for the businessman is not so much what an entrepreneur knows, but what science says. The best summary available in print was written by Fred Walters for Phil Callahan's *Ancient Mysteries, Modern Visions:*

"Science, perhaps to an extent unparalleled in any other field of human endeavor, has a very peculiar set of standards, norms, expectations, dogma and even rules. For instance, freshman science students are repeatedly hammered with the philosophy of the scientific method. The scientific method, a deductive form of reasoning, was designed to provide science with a foundation and framework into which all of the assorted bits of information could fit to form an integrated area of knowledge. It reaches not only into the cataloging and collecting of bits of information, but into actual discovery. It is also drawn upon by scientists to provide a logical way of finding the answers to problems.

"The first step in applying the scientific method to the solution of a problem involves carrying out a series of experiments designed to gather all facts about the particular problem being investigated. Then a simple generalization is formulated to correlate these facts. If successful, this generalization becomes scientific law. A jump such as this is seldom made, however, without an intermediate stage – the hypothesis. This is, for the lack of a better term, an educated guess. It is one idea that may serve to join the various facts observed. An hypothesis will be subjected to further experimentation in the attempt to find a flaw. If generally unrefuted, the educated guess will earn the status of theory, where it will likely remain for fear someone will find an exception. It is far more acceptable to disprove a theory than a law.

"There is a second scientific method that, although unwritten, has far greater impact on scientists and their findings. This is the reality of project funding, peer review and the publishing of scientific papers. These subjects were discussed in the scandalous book *The Double Helix,* by James D. Watson, one of the discoverers of the structure of the DNA molecule. He rocked the scientific world by discussing the behind-the-scenes power plays and jealousy and fighting for funding. But Watson offered these as an aside, showing that scientific discovery is a very human process, not a cold, mechanical ordeal filled with test tubes and microscopes. Discovery relies on a vision.

"Like Watson, Phil Callahan has not let the various bureaucracies and administrative tangles taint his love for science and life. Callahan left Louisiana State University at Baton Rouge because he wanted to study biological methods for insect control, while the system told him to study pesticides. Many of the discoveries explained in *Ancient Mysteries, Modern Visions* are still being bandied about by the scientific community. In fact, Callahan actually has a letter that states, 'You went too far,' implying that he discovered too much."

This code must be kept in mind when we recount the validating experiments contained in *Ocean Agriculture Hydroponics Project,* by Edwin M. Ederhamm III and Rhonda Holtzclaw (Florida Gulf Coast University College of Arts and Sciences, Division of Ecological Studies, 2001).

The experimenters used reconstituted ocean water enriched with nitrogen as their nutritional solution and compared their results to those obtained using a conventional hydroponic solution. The experiment grew 20 tomato plants in each of four raised gravel beds. The final report stated two beds were fed with the experimental solution, two with the solutions used by commercial hydroponic growers. Four variables were examined, this in order to compare experimental and control bed results with reference to growth, infestation by insects, quantity of production and product taste. The report noted that in terms of height growth, there was no difference between test and control plants. Final calculations revealed more bulk production from the beds treated with ocean water. Further, insect infestation was very low in the ocean-water beds compared to the commercial control plots. The report noted that "the addressed damage from the insects" fell below the "slight damage" category. The proximity of controls to test plants, and the level of infestation due to proximity, was not noted. The report said that control plants produced significantly larger tomatoes, but the experimental plants produced more tomatoes. Total production in kilograms did not differ significantly between the experimental and control plots.

According to the report, the ocean-grown tomatoes tasted "slightly" superior.

Chemical analysis confirmed chlorophyll levels. Higher levels of amino acids were found in the experiment fruit, as well as an increase in nutrient value. Otherwise, the $10,000 grant devoted to securing the university's blessing added little to existing knowledge. To be sure, it was agreed that the economics favored ocean grown in terms of efficient water use and cost advantage.

A report of this type is never without interest. It teaches something about the scientific procedures Fred Walters wrote about, and it suggests some of the questions a farmer should ask as his crops mature. It also illustrates the fact that instrumentation needs to be in the six-figure cost area. The Florida researchers used simple LaMotte instruments to check post-irrigation efficiency for total dissolved solids tests, all these after each flooding of the fields.

Temperature, oxygen and pH levels helped flesh out the records. The variables included were growth, insect infestation, production and taste. The one thing the sponsors of this report were entitled to, but did not get, was a tally of nutrient uptake, the primary motive behind ocean water hydroponics in the first place. This can be excused, however, because Gulf Coast University was a brand new school and really was not equipped to do a first-class study. Almost by definition, a professional study requires a lot of money. For $100,000 we can prove anything," said a researcher at Iowa State University during the early days of *Acres U.S.A.*

If proof has to be sanctified, so be it! In a nation that does not protect its borders, it seems doubtful that the mundane tomato can be grown profitably in any medium – not when the competition flies in specialty crops from South America and halves the cost of tomato production by using virtual slave labor imported from Mexico and Central America.

It was this reality that caused Don Jansen to close down all operations for a time and go back to the drawing board. Over the next few years, he encountered the nutritional health movement. This led him to grass and grass juices, a beautiful idea the food monopolists couldn't co-opt or compete against.

12

The Forgiveness of Nature

With or without university studies, a salient fact emerged from the work of Maynard Murray and from the observations of dozens of former collaborators. It was simply that plants each had their own diet and their capacity to uptake nutrients. As pointed out in my book *Weeds, Control Without Poisons,* that's the function of each weed, to uptake a special nutrient. Sometimes this makes a weed toxic. Often it turns a weed into a medicinal herb.

Crude measurements used to say that ocean-grown tomatoes picked up 40 nutrients, but Maynard Murray's more definitive test named 56 nutrients. As noted before, the sweet potato did a magnificent uptake job. It captured no less than 70 of the elements available in ocean water. Wheat grass of the type that Ann Wigmore prescribed for patients suffering all sorts of digestive metabolic diseases did a near-perfect job picking up nutrients when grown in good organic soil. In ocean water gravel beds, that same grass picked up close to 90 trace nutrients.

This fact stayed with Don Jansen and his associates. Wheat grass had a history even if it did not have much of a presence. Ann Wigmore spent her childhood learning about wheat grass and rye grass. During World War II she stayed in an isolated hideout with her mother and her grandmother, because all the men were off to war. As a consequence, the women had no food, no means of support, and no protection from raping and pillaging Russians. Grandmother would tell Ann to harvest grass off the lawn to keep them from starving. They ate the grass. This was not the diet they would have liked, but they never got sick. In fact, they were healthy throughout the war, even after spending weeks in their dugout. They made it through the war.

Ann Wigmore managed to come to the United States as a teenager under the sponsorship of an uncle who owned a dairy. Her job was to deliver milk house to house. As a youngster she drove a milk delivery cart drawn by horses. One day the cart tipped over. Ann broke both ankles. The uncle was terribly upset. Here he had paid for Ann's transportation, and now she was of no use and a burden. She had to stay at home, of course. Her ankles became swollen; gangrene set in. The feet turned black. It seemed obvious she would never walk again. She revealed how her grandmother told her that grass would save her life. Each day she asked to be set out in the yard during the day. She spent the day that way, plucking grass and eating everything within reach. Each day she asked to be moved this way or that for shade or sun, this to get at new grass. Over a period of time her ankles healed. She did not go to a doctor because she could not afford it.

She walked again, and she knew there was something about grass. In time she set up four centers called the Hippocratic Health Institute: one in Boston, one in West Palm Beach, one in San Diego, and one in Puerto Rico. Healing with wheat grass was the name of the game, and grass was the key. She told her story in a book, *Why Suffer?*

A plant can't take up a mineral if it isn't there, unless by chance it's in the air and encounters the stomata of a leaf or blade of grass. That is why Ann Wigmore insisted on good bioactive soil. That is also why open-pollinated corn will not grow well on soils that support hybrids. It is also why weeds will not proliferate in healthy bioactive soils.

It was suggested by Dr. Charles Schnabel, the "Sputnik grass expert," that his grasses contained 407 proteins when properly harvested. His long years of research validated Ann Wigmore's claim that sprouts were the greatest deliverers when made part of the meal. It wasn't grass alone that carried Ann Wigmore into the nation's headlines. She became the champion of raw foods, basing her hypothesis on experiments that challenged conventional science and then found validation as other scientists in other rooms confirmed her findings. She used the camera, the writing pad and narrative to make her points. Often the pictures—most of which were up in flames in the fire that took her life—best told the story.

One picture revealed how one chick grew rapidly and well, and how another wasted away. The one that grew rapidly had been fed uncooked food. The other chick got cooked food. The same experiment was carried out with cats with a protocol reminiscent of the famous Pottinger's cat studies. Again and again she proved her thesis — that enzymes are destroyed by heat and that minerals in far-ranging numbers are essential for good health. Further, she introduced the fact that the body well nourished has greater mental acuity.

"I carried out experiments with hamsters," she told this writer, "I used cooked and uncooked foods and certain sprouts and grasses. These can triple life spans."

She also described a monkey that was dying of cancer. Balanced nutrition not only saved the monkey, it brought the primate back to health.

There is incredible healing power in plants that take up a full inventory of elements.

Don Jansen accepted Ann Wigmore's findings wholeheartedly, and then improved on her work. He grew wheat grass in ocean water, ran his hydroponic farm until economics and public policy made it impossible.

At the end of World War II, Congress passed the Employment Act of 1946. This legislation set up the Council of Economic Advisors and ordered a structural balance for the United States, meaning parity for agriculture, perishable commodities included. *The Economic Report of the President* became a reporting medium for the state of the nation. Unfortunately, the political scene shifted, and in 1948 Congress passed legislation that literally demolished the goal of a stabilized economy.

In 1948 President Truman signed an executive order setting up GATT, which has now morphed into the World Trade Organization. For 10 administrations the national goal has been to lower commodity prices, consolidate farms, import food, export high-tech equipment, and relocate factories to poverty pockets overseas. GATT was enlarged in 1993 with the fast-track passage of NAFTA. Growing vegetables in ocean water to compete with cheap Mexican tomatoes and Chilean products no longer made economic sense. Jansen folded the operation and took a sabbatical pending the arrival of a new drawing-board product, one the mega-operations couldn't duplicate or compete against.

The human mind has its mysterious ways. Jansen revealed his days as a Nebraska rancher. He and his family raised cattle and also a field full of buffalo. The buffalo, he now reveals, is a cantankerous animal. No fence can hold a stampeding buffalo, and yet the bison rarely challenges a fence. But the animal likes its privacy. When rubbernecks point their cameras at the animal, it retreats to the far end of the place.

Most important, the buffalo likes grass. In fact, it consumes little else. That much was evident one day when Jansen tried to feed them alfalfa hay. The stack remained untouched. But when Jansen sprinkled ocean solids over a portion of the pasture near the usually shunned road, the buffalo mowed the grass as clean as any lawnmower. Moreover, they chose the treated area for a resting place. This confirmation of Maynard Murray's little book became coupled to the lessons Ann Wigmore had installed in the theater of Jansen's mind.

The growth of wheat grass for juicing became a hobby, an avocation. The market offered commercial juicers that could crank out grass juice in a volume suitable to a cottage industry. Something was missing in the equation.

The product could last some eight days. This cancelled out the possibility of economic imports. But the volume was so perishable it couldn't service much more than a neighborhood, much less a community.

Serendipity came to the rescue. Many Canadians resettle in Florida, having lost their taste for Canadian winters. One farmer visitor told Jansen about an exhibition and contest horse that was taking all the prize money at fairs and rodeos, even at the Calgary Stampede. The horse was named Big Ben. When the subject of wheat grass came up, mention was made of an automated machine called Opti-Grow. As the name indicates, it grew grass for Big Ben, the jumping horse, on a daily basis. This machine followed the world-champion jumper because the animal preformed best on fresh feed. In fact, any sudden switch to hay and grain was an insurance policy for failure. Wherever the horse went, a trailer with growing wheat grass followed.

Opti-Grow confirmed that all competition horses suffer badly and even get stomach ulcers when moved from fresh pasture to dry oats or hay. The show horse is a high-ticket property and merits care few human beings enjoy. The pastures for Kentucky racing stock portrayed in Kevin Conley's classic book *Stud* illustrates the point for those who wish to further examine the subject. The switch from grass to dry feed is now being tested at Doomben Turf Club in Australia.

It was the automated machine that caught Jansen's attention. Tracking the lead invoked the use of Internet search engines and diligence that would have tested the patience of Job.

"We got lucky," intoned John Hartman, one of Jansen's associates. It turned out that one Rob Smith, an engineer, developed Zero Grass, now Opti-Grow, soon to be Ethtecs. The original module migrated to Canada, where it helped Big Ben become "the winningest horse," John Hartman reported. Big Ben is gone now, but a competition honoring the big horse stays on.

The nutritional value and economics of pasture grass are discussed in *Reproduction and Animal Health,* by myself and Gearld Fry, and there can be no doubt that the same general concepts pertain to human beings.

If this book reads like a paean to grass, a few quotes from a real paean are in order. It was written by John J. Ingalls, a Kansas State Senator from 1873 to 1891. His speech *In Praise of Blue Grass* was published in *Kansas* magazine in 1872 and reprinted in *Grass: The Yearbook of Agriculture, 1948.*

"Blue Grass, unknown in Eden, the final triumph of nature, reserved to compensate her favorite offspring in the new Paradise of Kansas for the loss of the old upon the banks of the Tigris and Euphrates. Next to the divine profusion of water, light and air, those three great physical facts which render existence possible may be reckoned with the universal beneficence of grass. . . .

"Grass is the most widely distributed of all vegetable beings, and is at once the type of one life and the emblem of our mortality. Lying in the sunshine among the buttercups and dandelions of May, scarcely higher in intelligence than the minute tenants of that mimic wilderness, our earliest recollections are of grass.

"And when the fitful fever is ended and the foolish wrangle of the market and forum is closed, grass heals over the scars which are descent into the bosom of the earth has made, and the carpet of the infant becomes the blanket of the dead."

Grass has been the favorite symbol of the novelist, the chosen throne of the philosopher. *All flesh is grass,* said the prophet. *My days are as the grass,* said the troubled patriarch. And the persuasive Nebuchadnezzer in his penitential mood exceeded even those, and as the sacred historian informs us, did eat grass like an ox.

"Grass is the forgiveness of nature, the constant benediction."

John J. Ingalls could not have discovered the nutritional value of ocean water. But he understood the role of grass in providing healthy meat protein. Duplicating the wheat and rye grass that virgin soil once produced thus became the objective and reason for being of Don Jansen's reentry into the hydroponics he set aside in the wake of NAFTA.

There are always ideas for which the time has come. John Hartman's father was an ambassador to the former Soviet Union. To be a good ambassador, you have to exhibit a willingness to be bored.

It was during a field trip into the Soviet outback that the hosts unveiled a trailer that disgorged fodder out of one end, this in the middle 1980s. It seems a consulting firm named

Brown and Root (now a part of Halliburton) had international contracts of scope even in those days.

It was that bit of information, Ann Wigmore, a paean to grass and an Internet survey that took Hartman to Australia.

Australian entrepreneurs are marketing wheat grass juice fresh to consumers much as milk used to be delivered to the doorstep in America. Small units grow a ton a day in their trays. Large units produce 12 tons a day.

Total nutrition is a hefty goal. It happens also to be the one enterprise Walmart can't model, since even refrigerated wheat grass juice has a short shelf life, one that can't submit to long transportation rounds and retail handling.

It is axiomatic that the body will heal itself if it has a chance. Full-spectrum nutrition performs miracles.

"My father drank eight ounces of ocean-grown wheat grass a day during his last few years," Jansen now recalls. "Color returned to his hair. His eyes returned to 20/20, enabling him to discard his glasses. He received his driver's license at age 90. His skin took on a youthful appearance." After having toxicity-induced tumors all his life, he became cancer-free during his terminal years.

It takes only seven days to grow wheat grass suitable for juicing. Mankind has to opt for juice because, unlike the bovine, with its four stomachs, the human being cannot handle the strong fiber. It is the water of a plant that carries the nutrients. Cells do not consume fiber. They want liquid.

Researchers tell us that the fibers contain toxins. It is juice that enters the bloodstream. This has to be live juice. Norman Walker, the well-known juice man, invented a hydraulic press that could extract grass juice without destroying the enzymes. Usually a juicer develops heat that destroys the enzymes. Walker lived to age 119.

There are juice products that comply with the shelf life — read that shelf death — requirement. They are powdered, capsuled, preserved and chemicalized, far from the "forgiveness of nature." Jansen and associates maintain refrigeration and oxygen-sealed juice that will actually last 12 days without loss of qualities. Twelve days is stretching it a bit. A 10-day deadline is better. Delivered in a vacuum-sealed package, the nutrients and other carriers stay fresh that long.

Wheat grass juices are superb even when produced conventionally. As grown in ocean water, they trump all other growth mediums for a reason. Edward O. Wilson defines the scope of life on Earth in his books on ecology. He argues convincingly that there is no place in the world that does not play host to life, much of it microscopic, from Mount Everest to the depths of the Marianas Trench. The ocean houses the most fecund life of all. It contains five times more aerobic bacteria per square inch than any inch of soil on *terra firma*. That is why ocean water increases aerobic action in pastures, row-crop soils and fruit groves. Each plant has its own index. Palm trees and cypresses like their ocean water neat, uncut, robust and natural. For other plants, it is the function of the farmer to experiment, experiment, experiment. When ocean water becomes available as a concentrate, it becomes the role of the supplier to furnish the index plant by plant. Soils also have a tolerance index, clay being the poorest, loam and sandy soils being both tolerant and fertile.

At the beginning of the last century, New Jersey researcher George H. Earp-Thomas proved that size — as in ocean minerals — governed the entry of minerals into the root hairs of plants. Nature has excused the micron-sized nutrients from making the journey. It is to the root what the eye of the needle is to the biblical traveler. All plants seem to have their pecking order, grass being the king of the hill. It uptakes more minerals than any plant tested by Maynard Murray. There is a nutritional equation at work in plants, animals and human beings, Jansen said. Full nutrition guards plants from the ravages of diseases. This nutrition factor also answers the retailer's demand for shelf life of fruits and vegetables. Grass is much more delicate. It has this Omega-3 content, the right minerals, and its anointed shelf life. Sandwiched between reality is demand and biological possibility — grass juice may be health's last chance.

The salt content of ocean water, once believed to be toxic to growth, is now seen as the engine for intake, drawing up nutrients like a wick. If so, the absence of salt cancels out uptake of trace minerals, leaving alive the mistaken belief that fertilization automatically means uptake.

Grass can survive on no nutrition or on total nutrition. Rain belt grass is often deficient to a point that cattle starve on it. The same can happen to the hydroponic grass producer not conversant with ocean-grown methods. Rainbelt grass achieves cosmetic beauty but lacks nutrition – unless! The unless is a glass of diluted ocean water. If eight ounces of juice a day can keep a human being healthy, then what are the prospects for a beef cow on ocean-dressed pasture?

13

A Pantheon of Minerals

Whence come these minerals of which we speak? Their origins in the bowels of the Earth have been noted. Some may have arrived from the heavens. Readouts from high-priced instruments tell us that ocean water contains the following elements, give or take a few, depending on location near ocean vents and extraction methods. These appear below with the element name, symbol, Mendeleyev location and mass.

Element Name	Symbol	Number	Mass
Hydrogen	H	1	1.00797
Helium	He	2	4.0026
Lithium	Li	3	6.939
Beryllium	Be	4	9.0122
Boron	B	5	10.811
Carbon	C	6	12.0112
Nitrogen	N	7	14.0067
Oxygen	O	8	15.9994
Fluorine	F	9	19
Neon	Ne	10	20.18
Sodium	Na	11	22.99
Magnesium	Mg	12	24.31

Element Name	Symbol	Number	Mass
Aluminum	Al	13	26.98
Silicon	Si	14	28.09
Phosphorus	P	15	30.97
Sulfur	S	16	32.07
Chlorine	Cl	17	35.45
Argon	Ar	18	39.95
Potassium	K	19	39.10
Calcium	Ca	20	40.08
Scandium	Sc	21	44.96
Titanium	Ti	22	47.88
Vanadium	V	23	50.94
Chromium	Cr	24	52.00
Manganese	Mn	25	54.94
Iron	Fe	26	55.85
Cobalt	Co	27	58.93
Nickel	Ni	28	58.71
Copper	Cu	29	63.55
Zinc	Zn	30	65.37
Gallium	Ga	31	69.72
Germanium	Ge	32	72.59
Arsenic	As	33	74.92
Selenium	Se	34	78.96
Bromine	Br	35	79.91
Krypton	Kr	36	83.8
Rubidium	Rb	37	85.47
Strontium	Sr	38	87.62
Yttrium	Y	39	88.91
Zirconium	Zr	40	91.22
Niobium	Nb	41	92.91
Molybdenum	Mo	42	95.94
Technetium	Tc	43	98.91
Ruthenium	Ru	44	101.07
Rhodium	Rh	45	102.91
Palladium	Pd	46	106.4
Silver	Ag	47	107.87
Cadmium	Cd	48	112.4

Element Name	Symbol	Number	Mass
Indium	In	49	114.82
Tin	Sn	50	118.69
Antimony	Sb	51	121.75
Tellurium	Te	52	127.6
Iodine	I	53	126.9
Xenon	Xe	54	131.3
Cesium	Cs	55	132.91
Barium	Ba	56	137.34
Lanthanum	La	57	138.91
Cerium	Ce	58	140.12
Praseodymium	Pr	59	140.91
Neodymium	Nd	60	144.24
Promethium	Pm	61	146.92
Samarium	Sm	62	150.35
Europium	Eu	63	151.96
Gadolinium	Gd	64	157.25
Terbium	Tb	65	158.92
Dysprosium	Dy	66	162.5
Holmium	Ho	67	164.93
Erbium	Er	68	167.26
Thulium	Tm	69	168.93
Ytterbium	Yb	70	173.04
Lutetium	Lu	71	174.97
Hafnium	Hf	72	178.49
Tantalum	Ta	73	180.95
Tungsten	W	74	183.85
Rhenium	Re	75	186.20
Osmium	Os	76	190.2
Iridium	Ir	77	192.2
Platinum	Pt	78	195.09
Gold	Au	79	196.97
Mercury	Hg	80	200.59
Tallium	Tl	81	204.37
Lead	Pb	82	207.19
Bismuth	Bi	83	208.98
Polonium	Po	84	210

Element Name	Symbol	Number	Mass
Astatine	At	85	210
Radon	Rn	86	222
Francium	Fr	87	223
Radium	Ra	88	226.03
Actinium	Ac	89	227.03
Thorium	Th	90	232.04
Protactinium	Pa	91	231.04
Uranium	U	92	238.03

We rely on paleontologists and archeologists to tell us what happened with the North American continent. One single event suggests recall before we move forward to place ocean minerals under the microscopic eye. About 55 million years ago a firey asteroid crashed into the shallow sea near what is now the Yucatan Peninsula. It had been traveling at perhaps 85,000 miles per hour, give or take, and lost its way for reasons only speculation can supply. The crash terminated the age of dinosaurs, literally leveled most of the continent, extinguished species, annihilated woodlands and prepared the way for mountains to rise, savannahs to form and, not least, for mineral dusts to be distributed worldwide. One that asks for our attention is beryllium. It can't be found on land except at depths that invite paleontologists and their digging tools. Scientists date their finds by the beryllium layer, which was uniformly distributed when the asteroid struck. Yet beryllium shows up in a readout of ocean water. Does it have a role in enzyme formation? If Dr. Maynard Murray is correct, all the elements figured, all have a role, all of them governed by the law of homeostasis even if in concentration they are quite toxic.

The asteroid that struck also shaped our future, as in science fiction. Picture the scene, if you will. A star appears in the heavens. It will not graze, but will punch a hole into the planet as deep as Everest is high. Some of the fragments of exploded rock return to the heavens and a new orbit. Before the ocean can cool the wound, dust bellows skyward to circle the globe. The asteroid itself was perhaps three times the size of Australia' Ayres Rock, the largest monolith on the planet. The rock that struck with the explosive force of 100 million megatons of high explosives brought the Mesozoic age to a close.

On November 19, 1998, *Nature* published Professor Frank Kilt's definitive proof. He had found a piece of the asteroid ore taken from the ocean floor. The sample still contained the chemical traces of a carbonaceous chondrite. These chemicals are so rare they rate attention as the most miniscule of traces in ocean water.

The point here is that everything on Earth finds its way into the nutritional center of gravity, the ocean.

The connection between enzymes and specific minerals has been made in only a few cases. The full inventory of knowledge awaits discovery.

For now it is enough to supply a few notes simply to make the point that a shortage or marked imbalance of trace nutrients means malnutrition, bacterial, fungal and viral attack, debilitation and the onset of degenerative metabolic diseases.

It is a shortage that best defines our situation. Earlier in this book I called attention to the inability of hybrids to pick up trace nutrients even if they are present in the soil. This problem is exacerbated by the fact that too often the traces simply are not there. Soil scientists can test in vain for cobalt, a trace nutrient generally farmed out and totally missing in almost all American soils. Yet cobalt is essential if brucellosis in cattle and undulant fever in human beings is to be prevented.

If you run your finger down the list, going vertically, you'll see chromium and vanadium. These are the keys to enzymes that determine the glucose tolerance. A deficiency of chromium has been implicated in low blood sugar, hyperglycemia and finally diabetes. There may be more to the story. Since about the end of World War II, many municipalities have added sodium fluoride in one form or another to the drinking water, this on the theory that fluorine strengthens the apatite in teeth. Fluoride is one of four halogens: fluorine, chlorine, bromine, and iodine. Fluorine trumps iodine, for which reason iodine often does not make it to the thyroid, and thyroxin is not produced. Without thyroxin, sugar metabolism becomes a non-event. This deficit in being able to handle sugar is exacerbated by a sugar overbalance in the diet that has increased from about five pounds per capita in the 1930s to 135 pounds per capita at the present time.

The chromium molecule is required to burn fat, and chromium is simply missing from the soil and the food supplements due to unavailability. The chromium molecule is also a demanded element in muscle construction. Both chromium and vanadium function badly as synthetics. They function best when delivered by plant life, especially by grass.

Sulfur is a nemesis of cancer. Sharks concentrate ocean sulfur in their bodies, which is why some entrepreneurs offer shark cartilage to consumers. There are problems with all the recognized major nutrients and their tendency to achieve excess status with relevant cures that are worse than the cause. Just the same, it should be pointed out that sulfur protects the myelin sheath over nerve endings. It is thus an insurance policy against multiple sclerosis, Parkinson's disease and even Lou Gherig's disease. Sulfur may be toxic, but as it appears in ocean water, it has no side effects and no taste. Sulfur supplements are compounds, always inorganic compounds. The side effects can be awesome. Sulfur as it arrives in grass is organic, totally digestible. Sulfur compounds put on restaurant salads and in wine often cause allergic reactions, as evidenced by ringing wet around the collar and on the forehead, even breathing difficulty. The sulfur served up by ocean-grown grass scavenges free radicals, blunts food allergies, assists the liver in producing bile, adjusts pH, and assists in the production of insulin, sugar metabolism.

There seems to be a pecking order to mineral utilization, one so complex science can only hint at nature's complexity. For instance, that sulfur mentioned earlier requires vitamin C for absorption. In turn, vitamin C demands copper, and copper asks for zinc. Much as elements work together in ocean water, they support each other in the warm-blooded body.

Se, Mendeleyev number 34, is selenium. That short measure of selenium delivered by ocean-grown grass may be the lifetime protection against cancer. It's an antioxidant. It traps unstable molecules and skips damage. It helps confer immunity to viruses when ingested in nature's prescribed amount. There is research that suggest protection from neurotoxins. The mechanism has been identified. Selenium is used by the body to construct an enzyme that detoxifies staphs and build immunity. Unfortunately, selenium is generally missing in row crop soils except in some Western regions where it appears in toxic over-loads.

Selenium is implicated in muscular dystrophy, myalgia, cystic fibrosis, irregular heartbeat, Lou Gherig's disease, Parkinson's disease, Alzheimer's disease, Sudden Death Syndrome and many other abnormalities, sickle cell anemia and cancer included. There's more, namely, the nature of fat metabolism. The food industry no longer likes butter. It wants shelf life and therefore uses synthetic fats that do not melt at body temperature. This single fact also defines similar compounds as rancid fats filled to the brim with free radicals. Selenium is best able to deal with the rancid fats that have come to infect – yes, infect – our diet and its overload of free radicals.

We can digress to identify role and function just the same. Suffice it to say that viruses often inhabit the human system, sheltered from the immune system, often staying on for an opportunity to perform mischief years later. Various viruses and bacteria bow only to minerals that deal with the problem. These minerals have to be organic in the strict meaning of the term. They have to have a carbon passenger, ergo water soluble and of a size that permits transport not only into plants, but into the hiding places perceived to be unreachable by medicines.

That trace of silver in ocean water interdicts the activity of a virus that weakens a cell and turns it anaerobic. The cancer cell is no longer aerobic and oxygen consuming. It has turned itself anaerobic and finally goes into wild proliferation. The virus isn't alone in effecting cancer mischief. Parasites figure, as do toxins and pH levels at variance with human requirements. That is why ocean silver and zinc are so effective in preserving health. The law of homeostasis has decreed that these minerals are to be excreted if not required.

Move down the list a bit and you'll encounter copper. This mineral annihilates all parasites and intestinal worms. Entire texts have been written about parasites, some of them essential, most of them not. According to Hulda Clark, fully 97 or 98 percent of the American people are loaded with immune system-debasing parasites that take for themselves nutrition basically needed for health. This nutrient is either deficient or missing in the boxed foods sold across grocery store counters. The texts tells us that a copper shortage is often implicated in weight gain, cancer, a raft of allergies, high blood pressure and, yes, weight loss. These little copper healing creatures sail in the river of food and defy detection because of their size and metabolic duplicity. The placental barrier saves infants from many distress factors, but it can be breached by an overload of farm chemicals, mercury, atomic fallout and even malnutrition. Research is always indicated, but the promoters of ocean-grown wheat or rye grass are probably well within their mark when they point to copper and the array of minerals in ocean water and ocean-grown grass.

Zinc's association with copper is too well known to permit delay in presenting these few notes.

Water, of course, is H_2O – hydrogen and oxygen. The mere mention of oxygen suggests ozone and serves up the medical definition that ozone is a poisonous gas with no known medical use. A distinction has to be made: Nature's ozone, like nature's oxygen, is pure as the driven snow and both safe and efficacious. Ozone produced by high-voltage machinery is a nitric oxide acid gas. Most commercial machines produce a harmful gas. Ocean water does not create nitrous oxide. This is merely an aside and a warning to those who seek shortcuts via machines, when the real shortcut is daily use of wheat or ryegrass juice, especially juice from plants grown with ocean water. Oxygen is absolutely necessary for digestion.

Silver is a trace mineral that rarely finds a plant list, simply because it isn't there, at least not in soils. Its role in stomping out infections has been recognized by food supplement suppliers and now enjoys a brisk trickle. Organic silver requires a carbon component not generally available in inorganic supplements. Mere mention of one nutrient does not extinguish the requirement for another. The efficacy of silver in combating *Candida albicans* does not rule out the even better efficiency of raw garlic for the same purpose.

There is a classic diagram, provided to me by the late Harvey Ashmead of Albion Laboratories, Clearfield, Utah:

All so-called major and minor nutrient elements are microflora in which efficiency is energetically coupled. Don't let the word frighten you. It simply means that overdosing with one growth factor will change the entire spectrum. An excess of nitrogen will cause potassium deficiency. In fact, every excess disturbs the microflora's activity, chiefly through nitrification and fixation. Interrelations work their way all through the life chain. Here, for instance, are the mineral interrelationships in animals:

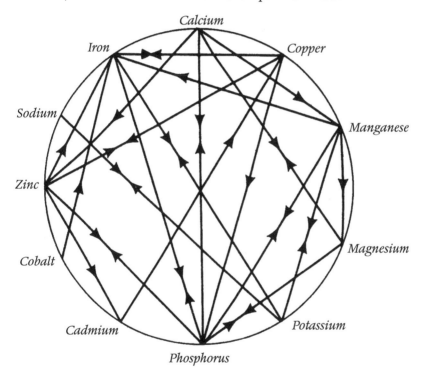

Based on the research of several investigators in animal testing, the above mineral interrelationships appear to be established. If a mineral has an arrow pointing to another mineral, it means deficiency of that mineral or interference with its metabolism may be caused by excesses of the mineral from whence the arrow originates.

The complexity of nature's arrangement seems awesome, a regular nightmare for the human being attempting to match wits, calibrate and supply one at a time. Here is where the ocean and its plenty come to the rescue.

You will note that fluoride is missing from the above list. Actually, there is no such thing as fluoride. There is a gas called fluorine. Combined with iron, they call it stanis fluoride, a compound. Combined with sodium, it becomes sodium fluoride. Both are said to assist the apatite crystals in teeth to harden. The idea is bogus and merely a device for unloading a waste product from the aluminum and phosphate industries into the water supply. The ocean does not construct these compounds and fluoride is not taken up by wheat grass grown in ocean water. The fluoride touted by dentists is a compound that turns stomach acids into fluoric acid. This particular acid is available in many grocery stores to take out rust stains in clothing. Sodium fluoride cancels out over 100 enzyme functions. The late John Yiamouyiannis attributed up to 50,000 deaths per annum by cancer to this contaminant.

The single factor that separates the useful from the useless is carbon. Carbon makes a mineral organic. The inorganic iron in processed foods is not easily assimilated. The worst-case scenario is hemochromostasis, a fatal disease, or iron supplement disease. Much the same is true when inorganic copper gets into the bloodstream, where it causes Wilson's disease, schizophrenia, the Jekyl-Hyde syndrome, enzyme shutdown and digestive failure. Copper and iron are not copper and iron, not if they are not organic. People often suffer aneurisms even though tested full of inorganic copper, this because of a copper shortage.

Even lead and mercury have their organic forms and arrive as harmless ingredients in plants. As heavy metals, they are among the most ubiquitous non-radioactive contaminants on planet Earth. Mercury in Portland cement and plastic is a hazard. That fog on the car window on a hot day is created by mercury escaping the plastic. It visits degenerative conditions too numerous to mention on mankind, and yet mercury and lead are listed as organic elements in the *CRC Handbook of Chemistry and Physics*. They are found naturally in plants and animals and ocean water, even though we are loathe to list them.

Confusion reigns supreme when human beings doctor themselves with compounds that pretend to supply missing nutrients. Calcium carbonate is a good horrible example. Calcium carbonate is simply one calcium, one carbon, three oxygens – better known as a blackboard chalk. It takes super-big activity to rescue this metabolic contaminant before the system can use the calcium. Usually it doesn't happen, and the chalk goes down the tube without any beneficial results. Muscle and leg cramps are a consequence of calcium depletion. Using blackboard chalk for a calcium source delivers osteoporosis. Simply stated, calcium carbonate is inorganic and not water soluble. Suffice it to say that most processed foods such as orange juice, cereals, etc., are loaded with this form of calcium. Ocean calcium is of a different stripe. It is perfect for plant assimilation, a crown jewel in the pantheon of essentials in the soil, the plant and the human being.

The business of counting elements – chromium picolanate, for instance–means making a complex molecular compound. This is the *modus operandi* for creating a new drug, many health food supplements, and all new drugs. The suggestion that the produce is delivering an element leaves stated the fact that side effects and reverse effect are always a legacy and frequently a debilitating consequence.

If this connection calls into question copper glutamate, zinc, pecolanate, vanadium picolanate and other complex molecular products, so be it. Blood vessels clogged by calcium are legion, as are triple and quadruple bypass surgeries because of blackboard chalk in the food supply and the absence of organic calcium in food crops.

Mere mention of these facts calls into question the recommended daily allowance (RDA). We ask and leave unanswered the question whether tests establishing RDA were accomplished with calcium carbonate or organic calcium!

The first element listed above is H, hydrogen. This is almost synonymous with pH, parts hydrogen. High hydrogen means acidity, or a low pH. Ascorbic acid equals hydrogen in a useable form. Too little hydrogen equals scurvy. Hydrogen is antagonistic to oxygen, leaves the latter element developing a shortfall of cellular oxygen.

A short digression may be in order. Grind grain, make it into bread, and you invite acidity. Let the grain sprout, then make bread, and the result is more alkalinity, a higher pH. Cooking food tends to lower pH because it destroys enzymes. As it declines, the ability of the body to absorb nutrients is diminished, leading to deficiency and disease. The pandemic of obesity and overweight, now an inescapable fact of American life, is a consequence of low pH in the food supply, among other factors. Viral diseases, cancer parasites, all gain permission for mischief from low-pH acidosis.

Cell division and blood clotting depend on minerals. They keep DNA and RNA activity at the cellular and subcellular level. They make vitamins possible. It is axiomatic that scientists can make vitamins, but they can't make minerals any more than they can make ocean water.

A full complement of minerals makes it possible for the body to self-regulate and self-repair its way out of most afflictions.

Linus Pauling, the only person so far to win two unshared Nobel prizes, once pointed out that you can trace every illness, every disease and every infection to some mineral deficiency. Any mineral deficiency always means there are even more mineral deficiencies waiting in the wings. It is equally true that most of the major degenerative diseases have been developed in test animals by withholding or manipulating critical trace minerals.

These minerals have been scoured from agriculture sites over the last two centuries just as surely as if they had been vacuumed out of a family room carpet.

The shocking absence of cobalt and chromium from New Jersey soils was recorded early last century by George H. Earp-Thomas. The issue of missing trace minerals and their role in plant and animal health consumed the working lifetime of William A. Albrecht at the University of Missouri. It also enriched the archives of Friends of the Land at Louis Bromfield's Malabar Farm in Ohio. Many of the great professors of the 1930s and 1940s amassed agronomic knowledge right up to 1949, when toxic rescue chemistry became established orthodoxy and agriculture was sent reeling into an uncertain world.

There is a mineral called molybdenum. Its function is to expunge waste from the body. Unfortunately, it is generally missing as though it went down under with beryllium when the asteroid collided with Earth. The only source seems to be ocean water.

Briefly, the anatomy of disease control and reversal of degenerative metabolic diseases is seated in the organic mineral diet and the vitamins controlled and dispensed by nutrients.

Thus, magnesium walks hand in hand with calcium. They go together like ham and eggs. The lack of one diminishes the role of the other.

None of these problems are easily solved with a handful of pills. All bow to all the minerals in the right form. Magnesium cancels out migraine headaches. This is merely an aside, a hint at the complexity of nature's demands and a recipe for meeting these demands. The pharmacy pretends to have drugs for asthma, anorexia, neuromuscular problems, depression, tremors, vertigo, organ calcification, etc., all when magnesium is the shortage. There is no need for calcium blockers or the alchemy of synthetic medication. The point here is that there is an absolute shortage of minerals in the food supply. The wheat grass juice that Ann Wigmore developed seems to be a final benediction and absolution for the transgressors of civilization.

There are mysteries in the ocean we hardly dare mention. Consider that 20 percent of the Earth's surface contains gold, organic gold. There isn't enough of it to justify setting up an extracting operation, but ocean water has enough of a trace to make a few suggestions. The literature suggests its offering in battling alcohol addiction, natural problems, circulatory problems – indeed a raft of anomalies that could fill this page. Its presence in ocean water is not a curse, rather a gift no less treasured than was that gold delivered by the magi. Gold's assent in achieving deep sleep is a staple in folk medicine, albeit one ratified by research and modern experience.

Platinum appears on the list that anoints the opening pages of this chapter. If anything, the presence of platinum in ocean water is even more fortuitous than its gold content. It figures in dealing with PMS, circulation, cancer. It enhances the ability to sleep and sparks daytime energy. Here again, ocean particle sizes contribute to efficiency as well as balance.

These few notes merely hint at the vast complexity contained in energy from the ocean. It has been reported that silver annihilates no less than 650 viruses. It does this because of the valence charge that surrounds resistant molecules when silver is present and able to assert itself. Even though silver kills viruses and anaerobic bacteria, it never harms the friendly fellows, the aerobic bacteria. It will be noted that the most effective burn ointments are silver-based.

A total of 90 elements have rated mention in this chapter. Others bask in silence. We do not know all the answers, or even the questions: Henry Schroeder, in writing *The Trace Elements and Man,* suggested another 400 years would be required to discern the role of each mineral if the present rate of discovery is maintained. Maynard Murray and Edward Howell calculated equal time for enzymes, knowledge of which is enlarged every day.

While we wait, the ocean abides and ocean-grown grass waits in the wings for those with the wit to use it.

Afterword

If anything, *Fertility from the Ocean Deep* is merely a preliminary report, a toe in the door, so to speak, of an investigative adventure that is bound to gather speed and move ahead during century 21. The physician Maynard Murray was a full-line producer and a serious scientist, both at the same time. It must have been a crushing burden to supervise field crops, hydroponic gardens and commercial vegetable production while at the same time seeing dozens of patients a day, finally supervising entire medical facilities in Florida. I am told that the good doctor read widely, consuming a book or two a night as well as the professional literature. Ed Heine, his Chicago-area based collaborator, assured me that Murray kept detailed and accurate records, most of which emerged in *Sea Energy Agriculture,* a pioneer attempt at setting down numbers, plots and results. Lee Comb of Leesbury, Florida, provided me with a narrative of his work and association with Maynard Murray, and, of course, Don Jansen of Fort Meyers, Florida, recovered his experiences for readers in an October 2003 *Acres U.S.A.* interview. Many Midwest farmers worked with Murray, and in each case the results, successes and failures became a part of the physician's record.

Unfortunately, Maynard Murray died prematurely.

His secretary, Cathy Jones, passed away within two years. The records were turned over to the estate, and now the trail grows cold.

I have been there before. While writing *Unforgiven,* I discovered a key figure in the parity story still alive at Sioux City, Iowa. I tripped north to find the cache of documents I needed to flesh out the story. Unfortunately, my contact had become senile, probably a victim of dementia praecox, and his conservator had thrown everything not needed for the income tax into a trash hopper. Figuratively speaking, that is what happened to the life and work of Maynard Murray. The task of recovering the value has been challenging and rewarding. Generally speaking, I have given titles of books consulted when this seemed appropriate. The papers of Gulf Coast University have been examined and quoted, as have letters from Ed Heine and Lee McComb and Maynard Murray's own report to the 1976 Acres U.S.A. conference.

The body of this text suggests a new dawn and a new appreciation of the dilemma facing planet Earth.

When I started *Acres U.S.A.* in 1971, the planet had already reached its carrying capacity, this according to the research arms of almost all ecology organizations with international standing and credentials. The false premises that "swept the republics of learning," to quote Sir Albert Howard, were conserving the resources of soil, water and air at a rate quite unsustainable. There was a demand for an agriculture friendly to the ecology, conservative in terms of resources, imaginative in terms of innovation. Maynard Murray's quite simple effort to harness a seemingly inexhaustible resource, ocean water itself, is so disarmingly simple it cries out for consideration.

In addition to those already cited, a special note of appreciation is due to Lynn Quale for research assistance, Anna Ross for the help a scribe always needs, for Don Jansen's encouragement and John Hartman's personal ratification of the premises expressed herein.

Always a realist, Maynard Murray lent his pioneering spirit, his profession and his passion.

"My main interest is spreading this concept. At this point in my life, I don't feel a need to prove anything anymore. I've traveled all over the world in search of answers to the question of why sea mammals don't get cancer. I've traveled every sea coast of every continent in search of large deposits of sea solids. I'm more than ever convinced that seaponics can go a long way toward wiping out hunger in our world and in the process improve man's health.

"You know, one million people die of starvation every day. And the irony is, we have what it takes to feed everybody on Earth. It's sad to see kids in India with tears in their eyes, but they can't cry. They're just too weak.

"Now, I'm no do-gooder, don't get me wrong, but I do feel an obligation to help others. The few lives a doctor can save or prolong is commendable . . . but so much more could be done with the right diet. I just know it."

Sources

Sea Mineral Agriculture Product Suppliers

Ambrosia Technology (Sea-Crop)
P.O. Box 6, Redmond, WA 98577
(360) 942-5698 (phone)
Info@sea-crop.com
www.sea-crop.com

Creation Sea Mineral
39458 Highway HH, Rayville, MO 64084
(660) 352-6383 (phone)
creationseamineral@gmail.com
www.creationseamineral.com

Ocean Solution
14001 63rd Way North, Clearwater, FL 33760
(888) 674-7696 (phone)
info@osgrown.com
www.oceansolution.com

SeaAgri, Inc.
P.O. Box 88237, Dunwoody, GA 30356
(770) 361-7003 (phone)
(888) 868-6051 (fax)
contactus@seaagri.com
www.seaagri.com

SeaMineral.com Inc.
P.O. Box 3707, Grand Junction, CO 81502
(877) 835-5555 (phone)
(970) 245-7024 (fax)
info@seamineral.com
www.seamineral.com

Index

Acres U.S.A., 7, 17, 53, 89, 92
adenine, 64
agar, 99
Albrecht, William A., 4, 12, 31, 34, 162
alkaline, 56
aluminum, 7, 45
amalgam, 95
amylases, 74
Ancient Mysteries, Modern Visions, 137
anhydrous ammonia, 60
anion, 43
antioxidant, 70, 156
Antunes, Francisco, 117
arteriosclerosis, 14
Ashmead, Harvey, 159
atomic weight, 27
Avian luekosis, 58
Azotobacter, 57, 90

Backster, Clive, 67
bacteria, 98
Baja Peninsula, 23, 35, 53, 104
battery acid, 92, 96
Bear, Firman, 3
Béchamp, Antoine, 2, 6
Bertrand's Law, 45
beryllium, 154
Big Ben, 145
Bittner virus, 131

blood, 53, 55, 111, 122, 126, 133
blood plasma, 112, 115
blue-green algae, 88-89
boron, 44-46
brucellosis, 6, 31, 155
buffalo, 102-103, 105, 144

C3H mice, 58, 68-69, 131
cadmium, 12, 51
calcium, 8
calcium carbonate, 161
Callahan, Phil, 111, 137-139
cancer, 12-13, 27, 58, 63-64, 69, 108, 129, 131, 157
carbon, xi, 100, 109, 157, 160
carbon dioxide, fixing, 88
carbonaceous chondrite, 155
Carson, Rachel, 64
cation, 38, 43
cattle, 131
cell division, 124
center rot, 57, 127-128
chelation, 101-102
chickens, 40, 58, 132
chlorine, 45
chlorophyll, 7, 10, 54
chloroplasts, 31
chromium, 12, 51, 155-156
chromosomes, 9
citric acid, 98
Clark, Hulda, 158

Clostridium, 57
cobalt, 6, 12, 31 51, 155
compost, 21, 125-126
copper, 7, 27, 44, 157
corn, 25, 38, 40, 60
corn blight, 57, 60-61
corn rust, 57
corn smut, 57, 130
corn, hybrid, 5, 6, 26
corn, open-pollinated, 5, 26
cows, 60
Crick, Francis, vii
crop yields, 90, 96
curly leaf, 57, 127
cytosine, 64

deficiency, 43, 46
dilution, 59
distilled water, 115
Dixon, Bernard, 100
DNA, vii, 59, 64-66, 71
dog experiments, 112
Double Helix, The, 138
DuPont, 93-94

E. coli, 2
Earp-Thomas, George, 6, 31, 88,
     148, 162
Eco-Farm, An Acres U.S.A. Primer,
     73
ecology, 148
Ederhamm, Edwin M., III, 139
electrical energy, 9
electricity, 100
electrolyte, 6
electron, 27
enzyme action, 72
Enzyme Nutrition, 70
enzyme nutrition axiom, 71
enzyme, disruptors, 73
enzymes, 75-76
enzymes, digestive, 73
enzymes, food, 73
enzymes, metabolic, 73-74
excess, 43

Federal Register, 3-4
Federal Security Administration, 3
ferric chloride, 55

fluoride, 95
fluorine, 114, 155, 160
Franklin, Benjamin, 15
free radicals, 70-71, 156-157
Fry, Gearld, 145
fungi, 89
Funk, Casimir, 77

galls (smut), 40
gene manipulating, 90
gene therapy, 65
General Agreement on Tariffs and
     Trade (GATT), 136, 144
Gerson, Max, 114
glucose tolerance, 155
gold, 93, 163
grass, 52, 146, 148-149, 156
grass juices, 140
Great Salt Lake, 119-121
guanine, 64
Gulf Stream, 15

Halbleib, Ernest M., 5-6
halogens, 155
Heine, Ed, 2, 10, 24-25, 35, 38,
     49-51, 53, 68, 89
Heine, Ray, 35
hemochromatosis, 55, 160
hemp, 94
hens, 41
hibiscus plant, 67
Hippocratic Health Institute, 142
Holtzclaw, Rhonda, 139
homeostasis, 7, 16, 45, 154, 157
Hook, Robert, vii
horses, 58
How to Grow World Record
     Tomatoes, 97
Howard, Sir Albert, 5, 8
Howell, Edward, 70-71, 73-74
hydrogen, 161-162
hydrogen peroxide, 70, 99, 108
hydroponic, 56, 67-68, 91-92, 97,
     99, 132
hydroxyl, 70
hypothesis, 138

infectious organisms, 116
inoculations, 77
inorganic elements, 10, 87-88, 100
inorganic nutrients, 13, 54
inorganic salts, 55
insect problem, 121
insecticide, 120, 132
iodine, 45
ions, 31
iron, 44, 56
irradiation, 34, 71
isotonic water, 113, 115-116
isotopes, 27, 29-30

Jacques, Ben, 24-25
Jansen, Don, 52, 77, 100-101, 103-104, 108, 135-136

Kilt, Frank, 155

L'Eau de Mer, Milieu Organique (Ocean Water, Organic Matrix), 111-112
LaMotte instruments, 140
leaching, 124
lead, 92
lipases, 74
live food, 137
Lives of a Cell, The, 30
Look, Al, vii

magnesium, 8
Magnificent Microbes, The, 100
malnutrition, 12
manatees, 13
manganese, 16, 44, 58
McComb, Lee, 119
McCrary, Paul, 137
Mendeleyev, 151-154
Mendeleyev chart, 31
Mendeleyev Periodic Table,, 45
Mendeleyev, Dmitri I., 27
mercury, 92-95, 160
microorganisms, 109
microwave, 71
mineral deficiency, 162
mineral interrelationships, 159

Mineralization: Will It Reach You in Time?, 52
mitochondria, 30
modern farming, 34
molybdenum, 27, 31, 44, 163
mosaic virus, 57, 128
Murray, Maynard, xi, 1, 7, 12-13, 33, 38, 50-51, 53, 87, 104, 123, 129, 165-166

N, P and K, xi, 5, 26, 43, 56, 92, 97, 129
NAFTA, 144
National Nutrition Plan, 3
neutrons, 30
nitrogen, fixing, 57, 88
North American Free Trade Agreement (NAFTA), 136
Nutrition and Physical Degeneration, 13

oats, 39
Ocean Agriculture Hydroponics Project, 139
Ocean Energy Testament, 119
ocean plasma, 113-114
ocean salts, 24
ocean solids, 25, 37, 38, 49-51, 58, 90, 97, 102-103, 121-124, 128, 132
ocean water, 53, 113, 122
ocean water therapy, 112, 117
Omega-3, 148
Opti-Grow, 145
organic compounds, 88
organic elements, 54
organic materials, 10
ostrich egg, 10
oxidants, 70
oxygen, 16, 70-71, 109
ozone, 158

parasites, 10, 158
parity, 106-107, 143
Pasteur, Louis, 2, 75
pasteurization, 75
Pauling, Linus, 162
peach trees, 127
pepsin, 77

Periodic Chart of Elements, 27
Peru, 52
pesticides, 88
pH, 91-92, 96, 98, 161-162
phosphorus, 56, 66
pig physiology, 61
pigs, 42
Pilsworth, Alwyne, xi
platinum, 164
potassium, 8
potassium chloride, 55
Price, Weston A., 13
proteases, 74
proton, 27, 30

Quinton, René 111-116

rabbits, 59
radiation, 64
radiomimetic, 9, 34
rate of application, 25
rats, 58, 131
raw foods, 143
recommended daily allowance
    (RDA), 161
renal disease, 93
Reproduction and Animal Health,
    145
rhizobial bacteria, 30
Roosevelt, Franklin D., 107
roosters, 41
rusts, 26, 39, 130

salmon, 63
salt fertilizers, 26
Sargasso Sea, 15, 52
Savage, Albert Carter, 52
Schnabel, Charles, 142
Schroeder, Henry A., 30, 69-70,
    164
*Science Society Proceedings of
    1948*, 3
scientific method, 138
Sea Energy Agriculture, 37-38, 53,
    102, 123, 126
Sea of Cortez, 62
sea solids, 18, 39, 44
sea solids, application, 60
sea solids, effects of, 62

seals, 13
Seaponics, 68, 104, 130
seaweed, 12-13
secondary trace elements, 45
Secretary of Agriculture, 106
selenium, 51, 156-157
selenium, excess, 62
shark cartilage, 17, 156
shelf-life, 17
Silent Spring, 64
silicon, 45
silver, 157-158, 164
Simpson, Harold, 18, 69
sodium, 8, 45
sodium chloride, 53-54, 56, 122,
    126, 133
sodium fluoride, 160
soil fertility, 125
soil matrix, 87
Soil, Grass and Cancer, 7, 27
soybean, 76
Sparactolambda looki, vii
spreader, sea solids, 35-36
Stabilization Act of 1942, 107
staphylococcus bacteria, 127
Steagall Amendment, 107
Steinbrown, Adolph, 5
sulfur, 156
superoxides, 70-71

teeth, 155, 160
Terapia Ortomolecular Natural,
    117
tests, lab, 125-126
theory, 138
therapeutic, 18
Thomas, Lewis, 30
Thomas, Luther, 92
thymine, 64
thyroid, 155
tin, 12, 51
tobacco mosaic virus, 128
tomatoes, 33, 75, 91, 96, 128, 139
toxic rescue chemistry, 8, 26, 34,
    44, 107, 163
trace element, 23, 43-44
Trace Elements and Man, The, 30,
    69, 164
trace mineral keys, 70

trace minerals, 6-7, 34
trace nutrients, 4-6, 31
trace nutrients, imbalance, 155
trout, 13
tuna, 16-17

undulant fever, 6, 31, 155
United States Department of
    Agriculture, 3, 103
Utah Salt Flats, 18

Valentine, Tom, 37
virus, 10
Voisin, André, 7, 11, 27

Walker, Norman, 147
Walters, Fred, 137
Watson, James D., vii, 138
Weeds, Control without Poisons,
    141
Weinberg's Principle, 45

whales, 13-14, 63, 76, 109, 129
wheat, 62
wheat grass, 18, 75, 141, 143-145,
    147-148, 158, 160
wheat grass juice, 163
Whitmer, Ted, 36, 37
Why Suffer?, 142
Wigmore, Ann, 77, 141-143
Wilber, Charles H., 97
Wilson, Edward O., 148
Wilson's disease, 55, 160
world trade, 136
World Trade Organization, 136,
    144

X-ray photography, 64, 65

Yiamouyiannis, John, 160
Yucatan Peninsula, 154

zinc, 7, 44

Also from Acres U.S.A.

Sea Energy Agriculture

BY MAYNARD MURRAY, M.D.

Reprinted by popular demand! Maynard Murray was a medical doctor who researched the crucial importance or minerals — especially trace elements — to plants and animals. Beginning in 1938 and continuing through the 1950s, Dr. Murray used sea solids — mineral salts remaining after water is evaporated from ocean water — as fertilizer on a variety of vegetables, fruits and grains. His extensive experiments demonstrated repeatedly and conclusively that plants fertilized with sea solids and animals fed sea-solid-fertilized feeds grow stronger and more resistant to disease. *Sea Energy in Agriculture* recounts Murray's experiments and presents his astounding conclusions. The work of this eco-ag pioneer was largely ignored during his lifetime, and his book became a lost classic — out of print for over 25 years. Now this rare volume is available to a new generation of readers. *Softcover, 109 pages. ISBN 978-0-911311-70-9*

Eco-Farm: An Acres U.S.A. Primer

BY CHARLES WALTERS

In this book, eco-agriculture is explained — from the tiniest molecular building blocks to managing the soil — in terminology that not only makes the subject easy to learn, but vibrantly alive. Sections on N, P and K, cation exchange capacity, composting, Brix, soil life, and more! *Eco-Farm* truly delivers a complete education in soils, crops, and weed and insect control. This should be the first book read by everyone beginning in eco-agriculture . . . and the most shop-worn book on the shelf of the most experienced. *Softcover, 476 pages. ISBN 978-0-911311-74-7*

Weeds, Control Without Poisons

BY CHARLES WALTERS

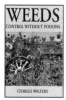

For a thorough understanding of the conditions that produce certain weeds, you simply can't find a better source than this one — certainly not one as entertaining, as full of anecdotes and homespun common sense. It contains a lifetime of collected wisdom that teaches us how to understand and thereby control the growth of countless weed species, as well as why there is an absolute necessity for a more holistic, eco-centered perspective in agriculture today. Contains specifics on a hundred weeds, why they grow, what soil conditions spur them on or stop them, what they say about your soil, and how to control them without the obscene presence of poisons, all cross-referenced by scientific and various common names, and a new pictorial glossary. *Softcover, 352 pages. ISBN 978-0-911311-58-7*

To order call 1-800-355-5313 or order online at www.acresusa.com

The Biological Farmer

*A Complete Guide to the Sustainable
& Profitable Biological System of Farming*

BY GARY F. ZIMMER

Biological farmers work with nature, feeding soil life, balancing soil minerals, and tilling soils with a purpose. The methods they apply involve a unique system of beliefs, observations and guidelines that result in increased production and profit. This practical how-to guide elucidates their methods and will help you make farming fun and profitable. *The Biological Farmer* is the farming consultant's bible. It schools the interested grower in methods of maintaining a balanced, healthy soil that promises greater productivity at lower costs, and it covers some of the pitfalls of conventional farming practices. Zimmer knows how to make responsible farming work. His extensive knowledge of biological farming and consulting experience come through in this complete, practical guide to making farming fun and profitable. *Softcover, 352 pages. ISBN 978-0-911311-62-4*

Hands-On Agronomy

BY NEAL KINSEY & CHARLES WALTERS

The soil is more than just a substrate that anchors crops in place. An ecologically balanced soil system is essential for maintaining healthy crops. This is a comprehensive manual on soil management. The "whats and whys" of micronutrients, earthworms, soil drainage, tilth, soil structure and organic matter are explained in detail. Kinsey shows us how working with the soil produces healthier crops with a higher yield. True hands-on advice that consultants charge thousands for every day. Revised, second edition. *Softcover, 352 pages. ISBN 978-0-911311-95-2*

Hands-On Agronomy Video Workshop DVD

Video Workshop

BY NEAL KINSEY

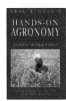

Neal Kinsey teaches a sophisticated, easy-to-live-with system of fertility management that focuses on balance, not merely quantity of fertility elements. It works in a variety of soils and crops, both conventional and organic. In sharp contrast to the current methods only using N-P-K and pH and viewing soil only as a physical support media for plants, the basis of all his teachings are to feed the soil, and let the soil feed the plant. The Albrecht system of soils is covered, along with how to properly test your soil and interpret the results. *PAL format also available, 80 minutes.*

To order call 1-800-355-5313 or order online at www.acresusa.com

The Non-Toxic Farming Handbook

BY PHILIP A. WHEELER, PH.D. & RONALD B. WARD

In this readable, easy-to-understand handbook the authors successfully integrate the diverse techniques and technologies of classical organic farming, Albrecht-style soil fertility balancing, Reams-method soil and plant testing and analysis, and other alternative technologies applicable to commercial-scale agriculture. By understanding all of the available non-toxic tools and when they are effective, you will be able to react to your specific situation and growing conditions. Covers fertility inputs, in-the-field testing, foliar feeding, and more. The result of a lifetime of eco-consulting. *Softcover, 236 pages. ISBN 978-0-911311-56-3*

Agriculture in Transition

BY DONALD L. SCHRIEFER

Now you can tap the source of many of agriculture's most popular progressive farming tools. Ideas now commonplace in the industry, such as "crop and soil weatherproofing," the "row support system," and the "tillage commandments," exemplify the practicality of the soil/root maintenance program that serves as the foundation for Schriefer's highly-successful "systems approach" farming. A veteran teacher, lecturer and writer, Schriefer's ideas are clear, straightforward, and practical. *Softcover, 238 pages. ISBN 978-0-911311-61-7*

From the Soil Up

BY DONALD L. SCHRIEFER

The farmer's role is to conduct the symphony of plants and soil. In this book, learn how to coax the most out of your plants by providing the best soil and removing all yield-limiting factors. Schriefer is best known for his "systems" approach to tillage and soil fertility, which is detailed here. Managing soil aeration, water, and residue decay are covered, as well as ridge planting systems, guidelines for cultivating row crops, and managing soil fertility. Develop your own soil fertility system for long-term productivity. *Softcover, 274 pages. ISBN 978-0-911311-63-1*

To order call 1-800-355-5313 or order online at www.acresusa.com

Natural Cattle Care

BY PAT COLEBY

Natural Cattle Care encompasses every facet of farm management, from the mineral components of the soils cattle graze over, to issues of fencing, shelter and feed regimens. *Natural Cattle Care* is a comprehensive analysis of farming techniques that keep the health of the animal in mind. Pat Coleby brings a wealth of animal husbandry experience to bear in this analysis of many serious problems of contemporary farming practices, focusing in particular on how poor soils lead to mineral-deficient plants and ailing farm animals. Coleby provides system-level solutions and specific remedies for optimizing cattle health and productivity. *Softcover, 198 pages. ISBN 978-0-911311-68-6*

Science in Agriculture

BY ARDEN B. ANDERSEN, PH.D., D.O.

By ignoring the truth, ag-chemical enthusiasts are able to claim that pesticides and herbicides are necessary to feed the world. But science points out that low-to-mediocre crop production, weed, disease, and insect pressures are all symptoms of nutritional imbalances and inadequacies in the soil. The progressive farmer who knows this can grow bountiful, disease- and pest-free commodities without the use of toxic chemicals. A concise recap of the main schools of thought that make up eco-agriculture — all clearly explained. Both farmer and professional consultant will benefit from this important work. *Softcover, 376 pages. ISBN 978-0-911311-35-8*

How To Grow World Record Tomatoes

BY CHARLES H. WILBER

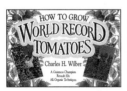

For most of his 80+ years, Charles Wilber has been learning how to work with nature. In this almost unbelievable book he tells his personal story and his philosophy and approach to gardening. Finally, this Guinness world record holder reveals for the first time how he grows record-breaking tomatoes and produce of every variety. Detailed step-by-step instructions teach you how to grow incredible tomatoes — and get award-winning results with all your garden, orchard, and field crops! Low-labor, organic, bio-intensive gardening at its best. *Softcover, 132 pages. ISBN 978-0-911311-57-0*

To order call 1-800-355-5313
or order online at www.acresusa.com